About the Author

Robin Hanbury-Tenison, OBE, DL, is the doyen of British explorers. A founder and president of Survival International, the global organisation supporting tribal peoples, he was one of the first campaigners to bring the plight of the rainforests to the world's attention. A veteran of over thirty expeditions, he is a gold medallist and one-time vice-president of the Royal Geographical Society. Among the films he has made for TV are some about the many long-distance rides he and his wife, Louella, have made in many countries. They live in a Domes-day manor on Bodmin Moor, where he has farmed for sixty years. His many books include *A Question of Survival*, *A Pattern of Peoples*, *Fragile Eden*, *Mulu: The Rainforest*, *The Oxford Book of Exploration*, *The Great Explorers*, *The Modern Explorers*, *Land of Eagles*, *Finding Eden* and his two autobiographies, *Worlds Apart* and *Worlds Within*.

Praise for *Taming The Four Horsemen*

'In my experience the world of environmental activism has no greater champion... This is a fabulous book because you won't agree with everything he says; he'll provoke you, seduce you, anger you but I promise he won't bore you. It's like pumping a mountain stream through your head. This book is activist Viagra and should be prescribed for anyone at risk of giving up thinking.'
– Sir Tim Smit

'This book could not come at a more appropriate moment, just when the world needs to think and decide whether we sink, dragging down with us all that we hold dear, or swim bravely, and sometimes uncomfortably, into a sustainable future. Matchless man: hugely important book.'
– Joanna Lumley

'A great champion of environmental activism and his extensive travels have given him many insights.'
– Sir Ranulph Fiennes

'As people start to get their heads around the full horror of today's climate emergency, we'll need a very robust conversation about potential solutions – including the highly controversial area of geoengineering. Robin Hanbury-Tenison has never been shy of that kind of controversy, which is always highly entertaining, whether you agree with him or not!'
– Sir Jonathon Porritt

Taming the Four Horsemen

Radical solutions to pandemics, war, famine and the death of the planet: an eschatological book

Robin Hanbury-Tenison

unbound

This edition first published in 2020

Unbound
6th Floor Mutual House, 70 Conduit Street, London W1S 2GF
www.unbound.com
All rights reserved

ISBN (eBook): 978-1-78965-110-2
ISBN (Paperback): 978-1-78965-109-6

Cover design by Mecob

Printed and bound in Great Britain by Clays Ltd, Elcograf S.p.A.

*Unbound is the world's first crowdfunding publisher,
established in 2011.*

*We believe that wonderful things can happen when
you clear a path for people who share a passion.
That's why we've built a platform that brings together
readers and authors to crowdfund books they believe in
– and give fresh ideas that don't fit the traditional
mould the chance they deserve.*

*This book is in your hands because readers made it
possible. Everyone who pledged their support is listed
at the front of the book and below. Join them by
visiting unbound.com and supporting a book today.*

Susan Adams-Cairns
Nicolene Akehurst
Michael Anderson
Rosemary Anderson
Andy Anderson
David Anderson
John Ashworth
Wesley Baker
Jenny Balfour-Paul
Joanna Barlow
Henry Barlow
Virginia Barneby
Elizabeth Barricklow
Niall Barton
Darcie Baylis
Loelia Baylis

Hokey Bennett-Jones
Martin Berthoud
Fiona Bicknell
Lucy Boutwood
Charlie Boutwood
Jason Brooke
Jennifer Bute
Henrietta Butler
Peter Cockburn
Pamela Codrington
Bill Colegrave
Gay Coley
Winchester College
Eton College
Nigel Collins
Nigel Collis

Julia Condon
Bryan Coode
Marcus Cotton
Ingrid Cranfield
Len Croney
Elspeth Crossley Cooke
Lucy Cubitt
Patrick Cunningham
Michael Cunningham
Felicity De Zulueta
Scilla Dillon
Feargal Duff
John Dyer
Martin Edwards
Peter Edwards
Louis Eliot
David Evans
Sarah Fairier
Odile Faull
Ranulph Fiennes
Sian Flynn
Liz Fortescue
Seymour Fortescue
Richard Forwood
Neil Foster
William Gore
Charles Granville
David & Lynn Gray
Jonina Gudmundsdottir
Sara Haboubi
Tim Hanbury
Louella Hanbury-Tenison
Jack Hanbury-Tenison
John Hare
Philippa Harkness
Nick Harrison
Ruth Hawley
Nick & Nicky Hayward
Helen Heslop
Julia Hicks
Brad Hicks
Charly Hicks
Carole Hill
Deborah Hinton
Crispin Holborow
Simon Holborow

Sarah Holcroft
Martin Holland
Jonathan Holley
RGS, Hong Kong
Mark Hubbard
Lucy Hyslop
Tim Izzett
Ana Jackson
Joan Jeavons
Robin Johnson
Jennie Keeler
Dan Kieran
Eleanor Koopman
Raghu Kumar
Sophia Ladanyi
Vanessa Leslie
Tasso Leventis
Anna Lewington
Megan Lewis
Paula Lewis
Phil and Rosie Leworthy
Nina Ludgate
Adam Ludlow
Edmund Ludlow
Ian MacWatt
Henry Madden
Debbie Manning
Robert Mansell
Barbara Martin
Rosa Mashiter
Ysenda Maxtone Graham
Colin May
Jill McCombie
John Medlock
Rosina Messer-Bennetts
Belinda Milbank
Andrew Mitchell
John Mitchinson
Antoinette Moat
Nicholas Money
Esme Moszynska
Stefan Moszynski
Marie Moszynski
Carlo Navato
John New
John Newey

Jeremy Nichols
Richard Nightingale
Thos O'Brien
James O'Fee
Sir Christopher Ondaatje
John Osborne
Graham Ovenden
Robert Owen
Anthony Page
Lord Palumbo
Ashley Parry Jones
Peter Partington
Jocelyn Phillips
Justin Pollard
Henry Poole
Jonathon Porritt
Martin Pumphrey
Philip Rambow
Julia Rhodes-Journeay
Margaret Ricardo
Graham Richards
Janet Richardson
Mike Richardson
Colin Ridler
Emily Roughton
Christopher Round
Joe Roxborough
Cathy Rozel Farnworth
Tony Ryde
Antony Sainthill
Leslie Schrock
Nick Schrock

Christopher Senn
David Sills
Tim Smit
Nick Smith
William Sowerby
Guy Speedie
Anthony Stancomb
Belinda Stewart-Cox
Merlin Charles Lord Sudely
Edwina Sutcliffe
Yan Swiderski
Ian Swingland
Susan Tenison
Maggie Teuten
Peter Throssell
Colin Thubron
Tom Tolley
Sarah Tolley
William Trinick
Philippa Tyrwhitt-Drake
Andreas Ugland
Jeremy Varcoe
Richard Vogt
Claire Wallerstein
Wayne Webb
Peter Whybrow
Vanessa Wilkins
James Williams
Gary Williams
Dick Willis
John Wood
Dr Masud Zarasvand

This book is humbly dedicated to James Lovelock and Ian Graham in gratitude:

the one for giving us insight and hope for the future;

the other for devoting his life to protecting the Maya legacy.

This frontispiece has most generously been pledged by Jason Brooke, director of the Brooke Trust, giving me the opportunity to promote the charity I care about most.

Survival International, which I helped found in 1969 and of which I have been chairman from the start and am now president, has been a truly world-changing organisation, of which I am intensely proud. Against all the odds, we have grown into the global movement for tribal people. From the Amazon to the Kalahari, from the jungles of India to the Congo rainforest, our dedicated team work in partnership with tribal peoples to protect their lives and land. They have so much to teach us, not least from their sustainable lifestyles, which have been consistently undermined. They suffer racism, land theft, forced development and genocidal violence just because they live differently. It must stop.

– Robin Hanbury-Tenison

www.survivalinternational.org

'Forests precede civilisations and deserts follow them.'
– François-René de Chateaubriand

'The clock is ticking. The Four Horsemen will be
here in the blink of a geological eye.'
– *The Selfish Ape*, Nicholas Money, 2019

Contents

The green pioneers *175*

Foreword

By Stanley Johnson

Full disclosure: I've known Robin Hanbury-Tenison as a friend for over forty years, and I've known about him for far longer than that. Robin's achievements loom large in my mind's eye, like a row of statues on Easter Island.

This is a man who in the sixties founded Survival International, to protect so-called 'primitive' peoples and to defend their right not to be intruded upon by 'civilisation' against their will. This is a man who, decades ago, led the groundbreaking scientific expedition to Mulu in the heart of Borneo. This is a man who, together with his wife, has ridden on horseback the length of the Great Wall of China. This is a man who has been at the heart of the Green movement since its very beginning.

Robin's book is no less authoritative for being readable and informative. He criss-crossed the Amazon at the end of the 1950s, when the rainforest was twice the size it is today. He witnessed at first hand over the decades the havoc wreaked by oil palm plantations in South-East Asia. He continues to despair at our inability to address the root causes of all environmental problems, *viz*.: the unchecked growth of human populations and our unsustainable patterns of consumption.

Ironically, I wrote the last of these few words as my plane

came in to land in Delhi. For a while, it seemed as though we might be diverted to Jaipur or another nearby airport owing to an air pollution crisis affecting visibility. But the smog cleared just in time.

I picked up *The Times of India* in the arrivals hall. 'Atmos-Fear,' the headline reads. 'Delhi Victim of Sick Choke.' Phew!

Robin Hanbury-Tenison, in this enthralling book, believes there is still just a chance of avoiding the plane crash ahead. It may still be possible to 'tame' the Four Horsemen of the Apocalypse. I hope he's right.

Introduction: The Four Horsemen

Four Horsemen of the Apocalypse (1887) by Victor Vasnetsov, Glinka State Central Museum of Musical Culture, Moscow

The world is beset by innumerable problems today, many of them of our own making. At the same time, we are so obsessed with the pressures of our own daily lives that it is hard to see the bigger picture. We live in a dual state in which we half-recognise the hopelessness of the direction we are travelling in, yet still cling to the belief that because we are such a resourceful species we will somehow find solutions. There are some hopeful signs: there is less starvation in the world today than there has been for centuries; we are developing brilliant technologies which are transforming our energy demands and our impacts on the environment; and there are brave and good people in national and international organisations working towards peace and prosperity for all. But on balance most would agree that we are now skating on very thin ice in many areas; the Four Horsemen of the Apocalypse lurk increasingly visibly in the background. There are many interpretations of what the author of the *Book of Revelation* meant, and I am far from the first to apply his message to modern times. The Four Horsemen of the Apocalypse are seen as the harbingers of catastrophes which can only be prevented by truly transformative changes and I use them as a template to frame my thoughts.

This is how I see it: the world faces four major problems. All are potentially existential threats to the human race, but all have possible solutions. The first, represented by the White Horse and a rider with a bow and arrow, is Pestilence, and the threat of global pandemics. (This alternative interpretation to the more traditional biblical representation of the White Horse as Conquest has been favoured for more than a century, and so I am using it here.) The second, the Red Horse, where the rider carries a sword, is War, often the result of pressure from overpopulation. The third, the Black Horse, is Famine, with scales indicating the weighing of bread; I link this to deforestation and the use of fossil fuels. The fourth, the Pale Horse, is Death, expressed as a skull, which I link to climate change and pollution.

There is no denying that their portents feel contemporary and relevant today and yet in manifold different forms these Four Horsemen have stalked the world innumerable times in the past. It is instructive, therefore, to look back at the collapse of previous civilisations to see what lessons might be learned.

My travels have taken me to diverse remote and wonderful parts of the world where I have met a great many people whose lives and cultures are very different from those I was born into. And so, alongside my exploration of the current challenges the world faces, I concentrate on and draw parallels with a society that flourished in an environment I know well and whose living descendants are the heirs to one of the world's greatest cultures, to see what lessons might be learned. Nonetheless, little did I imagine when, a few years ago my wife Louella and I travelled to the rainforests once inhabited by the Maya, that my journey would lead me to a new obsession with ways to save the world. But that is what has happened and this book is the story of how I came to believe some surprising things.

Prologue: Setting the scene

My journey begins just before dawn on top of one of the largest and least visited of all the Maya temples. From up here the world is covered in mist; only the top of Temple II pokes above the clouds. Like a diminutive tropical island, it is a Maya pyramid surrounded by treetops floating on an endless off-white sea.

This is Calakmul, the capital of the Kingdom of the Snake, Kaan, once one of the most powerful of all Maya cities and the implacable rival of the far better known Tikal, sometimes just visible far to the south. Now partially excavated and restored, it is seldom visited, as it is far from anywhere and there are few facilities. We are allowed, not without some intense discussion with the solitary guard, to sling our hammocks between trees at the entrance and to rise in the dark to climb Temple I by torchlight. A rare treat.

As the sky lightens in the east, the mist lifts and dense rain-forest is revealed in all directions, right to the horizon. In the darkness below not a single light twinkles; only gentle undulations disturb the ocean of treetops reaching to the forested hills in the distance.

Gradually a sky to rival the ceiling of the Sistine Chapel

forms overhead. Wispy pink, white, yellow, orange, red and mauve streaks of clouds weave patterns on the eggshell-blue canvas. It is a sight to strike awe into the most cynical heart; one which countless generations of men, women and children have gazed at in wonder. It lasts for only a short time before an intensity begins to make itself felt in the centre of the display. The light is brighter here and the eye is drawn towards the action, as though to the rising of the curtain at the start of an opera. The sublime spectacle of the overture begins to fade and the play unfolds.

The huge red orb of the sun inches its way over the rim of the earth and suddenly there is life in the forest as a cacophony of birdsong greets the dawn. First, there are the gentle hoots of motmots and a high-pitched shriek or two from brown jays; then from other small, invisible birds: robins, solitaires and trogons. Flocks of hysterical small green parrots speed in formation above the canopy, screeching to each other that they are late and must go faster. Pairs of their larger cousins fly past more sedately, their conversation more formal. In the forest canopy, large numbers of oropendola squabble furiously, like angry washerwomen arguing over who gets the best spot at the well. From deep in the forest, other, invisible, unidentified calls lay regular emphatic stress on the fact that *this is my place, my tree, my fruit. Join me my brethren, but keep away all rivals.* An eagle scours the treetops on lazy wings and gives a high-pitched shriek.

When the booming starts, all else is silence for a moment at the shock of it. The resonance is so intense, so insistent, so overwhelming in its sheer loudness that none who hear it for the first time can fail to be astonished and to wonder what on earth it can possibly be. It is a sound like no other in the world, one which can be heard from five kilometres away through the densest forest. It seems to be speaking directly to the listener.

'Beware,' it seems to say, 'I am all powerful and I am threatening you with dreadful things if you dare to challenge me.' This is the stirring cry still sometimes heard at dawn and dusk, and even occasionally during the heat of the day, throughout the remaining tropical forests of the New World. When the world of the Maya was young, more than 3,000 years ago, it would have resonated everywhere and spoken to them of powerful spirits. Eventually the people cut their forests down, as they always do, to feed their burgeoning cities and civilisations, and they must have hunted the howler monkey almost to extinction. Perhaps they became as rare then as now.

The first rays of sunlight strike us where we sit on the top of Temple I. A moment later, as the mist drops away, Temple II is revealed in all its glory. The stonework is a soft, creamy white, which contrasts delightfully with the surrounding greenery. In its heyday it would have looked quite different, cleared of all vegetation and largely painted bright red, with huge, grotesque multicoloured masks on the platforms. Upright stone pillars, life-sized statues, or stelae, would have stood in the courtyards below, as in the great cities of Greece and Rome. They were designed to impress and fill the spectator with awe, some soaring up over seven metres and weighing thirty tons. Although the carving may be exquisite and intricate and some of the faces reveal flashes of ethereal beauty, these were not primarily designed, it seems, for aesthetic reasons, as with the Greeks and Romans; instead they were meant to intimidate by showing the power and mystery of the Maya god-kings and their priesthood. The figures emerging from the rock were supermen, magicians, guardians of mysteries beyond the comprehension of ordinary mortals.

Now another, lower, temple shows white among the surrounding greenery. Colours rapidly sharpen and patterns form in a million shades of green wherever we look. The hard grey stone of the limestone rock from which the temples are built softens to a delicate pinkish yellow in the early light. The breathtaking steepness of the many steps soaring towards the square top jars the memory into feeling again the thigh-wrenching effort of going up them fast, as I always try to do. I am not good at heights and I like to get climbs over quickly, without looking down.

The Maya destroyed this, the second largest rainforest in the Americas, only for it to be abandoned for a thousand years and allowed to regrow. Now it is under threat again and seems to me to be a parable of our times. I have lived and travelled with foresters, biologists, anthropologists and natural scientists of all persuasions, trying to understand and interpret their myster-ies. I have also travelled in many of the remotest parts of the world, sharing the lives of indigenous people who have a very different view of the world from those I grew up with. I am a dreamer and a storyteller, an explorer, which in my book is someone who tries to change the world.

And I am angry. Deep down I despair at what we have done to our planet in my lifetime and what we continue to do at an accelerating pace. Can we not see the wonder of our world and what a crime it is for us, the only species with the brain and brawn to do so, to destroy it? Maybe there is a lesson to be learned here, at Calakmul. Perhaps the Maya civilisation needn't have vanished. Perhaps, if they had known what we know now, they could have managed their environment bet-ter and survived; perhaps we can learn from their mistakes and find a way not to vanish in our turn. The Maya story covers the whole range of human aspiration, creativity and folly but leads inexorably to the downfall of its protagonists. Mankind has gone through these same cycles so many times in so many different parts of the world, only to rise again from the ashes of its own destruction and, poorer but little wiser, try to rebuild the world it has destroyed. We seem unable to learn and yet we go on trying. This may be our last chance and looking back at the mistakes of the past may give us some ideas about avoiding them in the future.

For most of the last 200,000 years or so since something caused our brains to change and we made the great leap to become human, with the ability to think logically rather than instinctively, we have been hunter-gatherers. I believe that in many ways we attained the highest level of our humanity during this period, developing societies as near to perfection as the human race can aspire to. Nonetheless, the farmers had certain advantages which were to triumph in the end.

Quite a few hunter-gatherer societies still exist in the world today and during my lifetime I have been privileged to get to know and even live for a time with several of them. I have found among them a self-confidence based on absolute self-sufficiency, which is seldom if ever found among 'civilised' people. Civilisation means living in cities and cities depend upon agriculture. This means working the surface of the land to produce more than it does naturally. I have been a farmer all my life and I always took encouragement from the saying of Jonathan Swift: 'Whoever could make two ears of corn, or two blades of grass, to grow... where only one grew before, would deserve better of mankind, and do more essential service to his country, than the whole race of politicians put together.'[1] But I have come to doubt the validity of this philosophy as I have watched the terrifying destruction of my world, much of it done in the name of increased agricultural production. I have seen the rainforests of Borneo, once the richest environment on earth, disappear and be replaced by virtually sterile oil palm plantations, while some of us protested from the side-

1. *Gulliver's Travels*, II.7.

lines, trying to make the politicians see sense. I have seen the great Amazon forests felled for cattle ranching, a patently less efficient use of that environment, but one which still stirs the hearts of many Brazilians as a manly activity.

Looking at the story of the Maya has made me rethink much I had taken for granted, including the idea that civilisations can endure indefinitely. Far from it: worlds end. All are obliterated at some stage, often after they have burgeoned for some five hundred years. In fact, Luke Kemp of the Centre for the Study of Existential Risk at the University of Cambridge has recently studied almost a hundred civilisations that have come and gone throughout history and found that their average lifespan is 336 years. He gives 'climatic change' as one of the principal reasons for their demise.[2]

Why? I believe this is because, by their very nature, they destroy the resource base upon which they depend. Edward Gibbon's *The History of the Decline and Fall of the Roman Empire*, published from 1776, may have been the first book to expound this theory, based on the premise that it was the declining agricultural yields resulting from overexploitation of the land and the resulting climate change which weakened the empire and started the rot. 'The decline of Rome,' as he put it, 'was the natural and inevitable effect of immoderate greatness.' He also believed that removing the 'immense woods' of Europe since classical times had improved the climate. In this way, he was one of the first to postulate the existence of global warming. Discussing why Europe was now less cold than descriptions of ancient Rome indicate it to have been before, he wrote: 'The modern improvements sufficiently explain the causes of

2. Luke Kemp, 'Are we on the road to civilisation collapse?', http://www.bbc.com/future/story/20190218-are-we-on-the-road-to-civilisation-collapse

the diminution of the cold. These immense woods have been gradually cleared, which intercepted from the earth the rays of the sun. The morasses have been drained, and, in proportion as the soil has been cultivated, the air has become more temperate.'

The same connections between man's overuse of the environment and economic collapse have been used to explain the undermining of the Mycenaean society in ancient Greece, the Native American Indians of the Great Plains, the Mali Empire in Africa and, of course, the culture on Easter Island, among many others. The *Epic of Gilgamesh*, the oldest surviving written story, inscribed on tablets in the third millennium BC, describes the destruction of the forests of the Middle East. Two thousand years before Christ there were vast cedar forests clothing the mountain slopes of what are now the desert regions of Syria, Jordan and Mesopotamia. 'So Gilgamesh felled the trees of the forests and Enkidu cleared their roots as far as the bank of Euphrates.'

In the case of the Maya and the Petén rainforest it is easy to see what happened. From about AD 200, when the greatest Maya empire arose, vast swathes of forest were cut down to create farming land and orchards to feed the growing cities. Land was terraced and artificial lakes and aqueducts were built to manage the scarce water. This all worked well and mighty, powerful nation states grew, vying with each other to build ever greater temples, some as large as those found anywhere. The Danta pyramid (the name means 'tapir' in Maya), which actually dates from an earlier Maya era and was built in 400 BC, is the largest prehistoric building on earth, having a greater volume than the Great Pyramid of Giza. A faint forested mound on the far distant horizon to the south, over the border in Guatemala, revealed its location to us as dawn broke at Calakmul.

At its peak, the Maya population may have approached 10 million. The Tang Dynasty, probably the supreme period in Chinese history, with a population of 80 million, was at its height; the vast Buddhist monument at Borobudur was being built in Java; Islam was spreading into Spain; and the area of Viking conquest was expanding; but during what was to become known as the Dark Ages in Europe, the most successful of all Mesoamerican societies, the Classic Maya period, flourished between AD 250 and 900.

Then things started to go wrong. As the most fertile land and then the poorer soils were overexploited to feed the growing population, wars broke out over territory. The climate, deprived of the life-giving rainforest, began to alter. There was desertification, hunger, conflict, epidemics and, unmistakably, climate change. Sound familiar? The Four Horsemen of the Apocalypse – Pestilence, War, Famine and Death – then appeared, as they have done time and again throughout history to destroy yet another civilisation, and in the end everything fell apart. By AD 900 up to 99 per cent of the Maya had died, killed each other or moved away. For a thousand years the rainforest was left virtually alone, unoccupied, with no people; the clouds returned and the rain fell and, most unusually, the forest regrew in its entirety, until it was barely distinguishable from what it had been before man had laid a hand or an obsidian axe on it. Now its destruction is being replayed, as migrants, illegal foresters and cattle ranchers move in. Worlds do end, not just for the Maya, but also potentially for our own modern civilisations round the world which, though varied, face the same massive global challenges. The difference today is that there is something we can do about it. The time has come for us to take control.

Part 1

THE WHITE HORSE: PESTILENCE

Part 1

THE WHITE HORSE: PESTILENCE

The problem: Pandemics

In the fourteenth century the Black Death arrived in Europe and killed about half the population of the continent and between 75 and 200 million people worldwide.

Spanish flu, in 1918, accounted for between 50 and 100 million people, many more than had died in the First World War, perhaps more than have died in any war.

Penicillin and other antibiotics to treat viruses like the ones responsible for Spanish flu have stemmed the tide and prevented a similar disaster happening at the end of the Second World War. Antibiotics and vaccines to fight viruses have transformed medicine and saved millions of lives. However, the rapid emergence of antibiotic-resistant bacteria is occurring worldwide. Many decades after the first patients were treated with antibiotics, bacterial infections have again become a threat. This crisis has come about as a result of the overuse and misuse of these medications, including in the food chain through administration to farm animals. The pharmaceutical industry is also to blame for failing to develop new drugs fast enough. They say it is because of reduced economic incentives and challenging regulatory requirements.

Every few years we see a new disease in humans: SARS,

AIDS, mad cow disease, Zika, to name just a few. Most recently, Ebola took West Africa by surprise, but so far has been contained, with great difficulty. Unlike Ebola, flu transmits easily and mutates, making the risk of a new flu pandemic more dangerous still. There are many nightmare scenarios in the offing. Imagine if Ebola was transmitted aerially; if H5N1 (avian flu) were as lethal to humans as it is to chickens; if HIV, already a combination of four viruses, were to mutate again. Pandemics are extremely difficult to predict and difficult to prepare for. They all start with a random event: a pathogen crossing to humans from another species or a mutation. The real danger is a brand new virus, probably a strain of influenza.

There is little doubt that we face a major global pandemic before long. A recent study found that 90 per cent of epidemiologists believe that there will be a pandemic within their children's or grandchildren's lifetimes, in which, by one estimation, a billion people will get sick, 165 million will die and there will be a global recession and depression.[1] As cities grow exponentially and international travel becomes faster and easier, the probability of another pandemic becomes a certainty. Whereas it used to take weeks or months or years to circumnavigate the globe, it now takes about a day. Someone infected with a virus or a bacterium only has to get on a plane to infect thousands more. Although enormous effort is being expended to prevent pandemics, the increasing resistance to antibiotics, due to their overuse, makes one a virtual certainty. We have had some recent scares, culminating with Ebola, but so far we have been able to contain them. Conventional options to continue doing so are running out and we need to look at another

1. Larry Brilliant, 'My wish: help me stop pandemics', https://www.ted.com/talks/larry_brilliant_wants_to_stop_pandemics

way. I believe microbes could be the answer. Let me tell you why.

The solution: Microbes

Big fleas have little fleas,
Upon their backs to bite 'em,
And little fleas have lesser fleas,
and so, ad infinitum.
  – 'The Siphonaptera' (nursery rhyme)

The vast world of microscopic existence is quite as staggering as the infinity of space: it has been called the majesty of life unseen. Our bodies are literally riddled with life. The human microbiome is made up of microbes, which are microscopic organisms such as bacteria, archaea, protozoa and fungi that reside in and on our bodies. Did you know, for example, that there are as many creatures on your body as there are people on earth? There are 100 trillion bacteria cells in your colon, whereas our own galaxy, the Milky Way, only has 400 billion stars. And those microbes are not only fed by us, they influence our physical and mental well-being, our moods, our health and even our lifespan. Nearly all our interest in biodiversity through wildlife documentaries and so on concerns the relatively few animals, plants and insects we can see, and yet there is a vastly greater, and in many ways more important, unseen world only now being revealed to us in its immensity by the microscope. Much of it is breathtakingly graceful, too, and it is all busy getting on with being alive, even though so small that ten thousand bacteria could be squeezed inside a full stop. Realising this turns on its head the way we see the world. As Nicholas Money, the author of an excellent book on the subject, *The Amoeba in the Room*, says: '…we have been misled by our brains to exaggerate the importance of elephants. We need to employ a little imagination to appreciate the amoeba, and this readjustment holds the only sensible answer to the question about the true meaning of life.'

The potential for medicine and nutrition from research into microbes is inestimable and far outweighs any possible benefit from space research, for example. Far more should be spent on finding out how our own world works and trying to put right what is going wrong with it, largely due to us, than on hoping there is some magical solution 'out there'. It sometimes seems as though those advocating space exploration have given up on our own extraordinarily precious planet and are desperately seeking a way out. The scale of activity by microbes, which is increasingly being revealed, really does defy our ability to absorb statistics. Trillions of microbes and quadrillions of viruses are multiplying all the time on your face and mine, on our hands, in our guts. With every breath we take, we are sending bacteria into the air at the rate of about 37 million per hour (so it turns out that we do actually carry an aura about with us, as Buddhists have believed for millennia).[1] With every gram of food we eat, we swallow about a million further microbes.

Research and exploration into this vast and extraordinary world is just as awe inspiring as the exploration of outer space.

---

1. Eric Gershon, 'With you in the room, bacteria counts spike', https://news.yale.edu/2012/03/28/you-room-bacteria-counts-spike

For example, the bacterial species living in navels were examined in a recent experiment. They were found to be microscopic jungles colonised to an extent far greater than anyone had suspected. From just sixty navels swabbed, 2,368 species of bacteria were found, of which 1,458 could be new to science.[2] Others had only previously been observed in deep-sea hydrothermal vents and ice caps: another clue towards understanding how incredibly interrelated all life on earth is and in ways we are barely beginning to understand. Research is just beginning into the ancient microbial life, some as much as 40,000 years old, to be found deep in glaciers. Scientists who have extracted the DNA from these microbes, which number in the trillions, are excited by the prospect that they may hold the key to finding new antibiotics, to which we are not resistant. A truly vast area of research is unfolding.

We tend to think that all microbes, bacteria and viruses are bad things, which we need to wash off ourselves and eliminate as far as possible. There is even a cleaning fluid that promises to 'get rid of all known germs'. But the vast majority are harmless and increasingly we are discovering that most are highly beneficial; we couldn't live without them and, indeed, they shape our lives and our moods. They work with our bodies to keep them balanced and healthy. A bit late in the day we are beginning to realise that not only should we not be striving so hard to eliminate the 'harmful' germs, but that doing so is actually threatening our very survival. Overuse of antibiotics is not only killing off far more good germs than the bad ones that they should be targeting, it is rapidly creating resistance, so that they no longer do the job they were created for.

---

2. Hulcr J., Latimer A.M., et al. (2012) 'A jungle in there: bacteria in belly buttons are highly diverse, but predictable. *PLoS ONE* 7(11): e47712. https://doi.org/10.1371/journal.pone.0047712

The modern era of antibiotics started with the discovery of penicillin by Sir Alexander Fleming in 1928 and they were first prescribed to treat serious infections in the 1940s. Penicillin was very successful in controlling bacterial infections among Second World War soldiers, but soon afterwards resistance to the drug became a problem. Already, as early as 1945, Fleming was raising the alarm regarding antibiotic overuse. He warned that the 'public will demand [the drug and]... then will begin an era... of abuses.'

New types of antibiotics have been discovered, developed and deployed time and again since, often from samples sent in from particular environments around the world. My favourite example is vancomycin, which was first isolated in 1953 from a soil sample collected from the interior jungles of Borneo by a missionary. It has been used very successfully since for the treatment of many infections, including meningitis and *Clostridium difficile*, one of the most dangerous bacteria, resistant to most antibiotics. But the evolution of microbial resistance to vancomycin is a growing problem, in particular within healthcare facilities and hospitals, and the race is on to find replacements. And that is what happens, again and again. Each time resistance develops so that today the antibiotic pipeline is drying up and bacterial infections are, once again, a serious threat. Every year, at least 700,000 people die worldwide from infections that no longer respond to antibiotics. This toll is forecast to balloon to many millions in the near future. If there is a pandemic and, as I have said, the majority of epidemiologists believe there will be one sooner rather than later, and we have no way of preventing or curing the infection, then the outcome will be truly devastating. In 2016, Ban Ki-moon, the then secretary-general of the United Nations, called drug resistance 'a fundamental, long-term threat to human health, sustainable food production and development.

It is not that it may happen in the future. It is a very present reality – in all parts of the world, in developing and developed countries; in rural and urban areas; in hospitals; on farms and in communities.[3]

In 2017, Dame Sally Davies, England's chief medical officer, announced that the 'antibiotics apocalypse' may already be upon us. She said that 50,000 people are dying every year in Europe and the US from infections that antibiotics have lost the power to treat, and she described the threat of the loss of antibiotics to the world as being on a par with terrorism and climate change.

The answer, which is dawning fast on the scientific community, is to discover which the good microbes are, find out how they work on our behalf and make them our friends, rather than our enemies. It sounds easy, but it is a massive task and it constitutes pure exploration. People are always asking me if there is anywhere left to explore and my answer is that we barely understand the world we live in yet; even though we may have been everywhere, this inward exploration is really exciting. There is a vast universe of minute creatures, of which most of us are utterly unaware, waiting to be discovered. Just as increasingly powerful telescopes are helping to reveal what may or may not exist in outer space, so stronger microscopes and other scientific means of examining the minutiae of life on earth are revealing just how much there is in there – not out there!

We are already benefitting from microbes in lots of ways, and we always have. Yogurt, a food produced from the bacterial fermentation of milk, was probably invented in Mesopotamia 6,000 years ago. It has many values, as do other

3. https://www.un.org/sg/en/content/sg/statement/2016–09–21/secretary-general-remarks-high-level-meeting-antimicrobial

probiotics, microorganisms that provide health benefits when consumed. These are now recognised as friendly bacteria, naturally present in our digestive systems, which can help boost our immune systems and promote healthy digestion. Supermarket shelves are full of them, but they are only one of many potential ways we could be using microbes. This is a step in the right direction, but I believe we are only scraping the tip of a giant iceberg of opportunity for using microbes to combat and cure innumerable diseases and to keep our bodies healthy.

Microbes occupy different habitats within our bodies. There they form communities, wage war on each other and play a vital role in influencing every aspect of our lives. No one fully understands how this happens yet. The majority occupy our intestines and yet they demonstrably have an impact on our brains. Many conditions, from cancer to HIV to Parkinson's disease, are linked to microbial disorders. It seems there may be a superhighway from the gut to the brain. Research is in its early stages, but it could lead to novel therapies for all sorts of diseases.

Just as we are beginning to realise the potential for using microbes in medicine, so farmers, who used to think of microbes as pests that were destructive to their crops and animals, are starting to recognise their benefits. Soil microbes are essential for decomposing organic matter and recycling old plant material. Of course, this has always been part of agriculture: it is, for example, the basis of silage making, where lactic acid bacteria facilitate fermentation. Today, biotechnology is a rapidly growing segment of biological science and it has huge potential. It has a wide variety of exciting applications for developing sustainable agriculture in the form of biofertilisers, bioherbicides and biopesticides, including fungal-based and viral-based bioinsecticides.

Repackaging beneficial bacteria and fungi for agriculture in the same way as has been done with human probiotics and then delivering them to plants to alter their microbiome in ways that will boost growth, increase resistance to drought, disease and pests, and reduce farmers' reliance on fertilisers and pesticides is changing farming fundamentally. Top international agroindustrial companies like BASF, Monsanto, Bayer CropScience, Syngenta and Arysta LifeScience are rushing into a market that analysts believe could more than double in value, to $4.5 billion soon.[4]

Microbes occur everywhere and so an examination of those which occupy the human body is only a minute part of the story, but it is the part which most intimately interests us. Cells are the basic units of all life, everywhere. All living things are composed of cells and depend on them to function normally. Most of us are dimly aware of the existence of our human body cells: bone cells, blood cells, stem cells, cancer cells, etc. They number in the trillions and come in all shapes and sizes. What I was until recently totally unaware of (and I'm sure I am not alone) is that these tiny organisms represent only about 10 per cent of all the microbial inhabitants of the body. Microbes are living things which reproduce in a variety of ways, sexually and asexually; they self-destruct when damaged or infected (the inability of some to do this can be the cause of cancer); and they contain our genetic material in the form of chromosomes. Everything we do and are depends on them.

We have only begun to learn about them with the development of electron microscopes, which have revolutionised our awareness of the infinite variety of small things. An immense world of fascinating variety has been revealed. I find looking into the intimate depths and cavities of these living things as exciting as considering the exploration of terrestrial habitats, like caves, for example – and remember, they are alive and influencing our every action and mood.

Some are free moving, like male sex cells, which have long tail-like projections to help them seek out and unite with the larger female sex cells. Others become sick and start to divide uncontrollably, perhaps as a result of chemicals or radiation, and turn into cancer cells.

These little cells are small enough, but the other 90 per cent of lives within us and on us are even smaller. The human microbiota consists of all the microbes in and on our bodies. They live in distinct places in our individual ecosystems and they perform important functions without which we would die. Without them we could not digest our food or absorb the nutrients on which we thrive. The vast majority of beneficial microbes we have are constantly fighting the pathogenic microbes, the bacteria, viruses and fungi that cause disease.

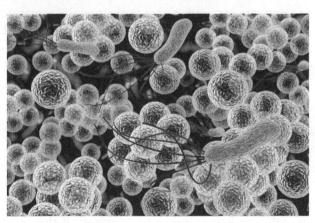

Gut microbes offer new worlds to explore

They even affect our moods and may be linked to how well we sleep. It is now generally accepted that what we eat and breathe has a wide and long-lasting effect on how we feel. The allergies and bad reactions that are increasingly becoming prevalent may well be caused by what is happening in our guts. Recent research is even linking our intestines and our brains in various disorders. Our brains, the most complex objects in the known universe, are now known to react directly to the state of our bacteria. Psychobiotics, cocktails of healthy bacteria, are now being considered as ways of boosting our mental health. This is opening up a whole new way of doing medicine. Our microbiomes may soon become the things our doctors look at first to assess our health.

Our microbes are supremely important to us in every imaginable way and yet our research into this infinity of life and matter is minuscule compared with the cost and effort expended on space research, for example. Instead of making plans to evacuate our own planet in the event of a man-made cataclysm we are doing all too little to prevent, we could use the astonishing tools nature has provided to make this world the paradise it has the potential to be.

MEDICINE

How could we use microbes to help avoid a pandemic and, indeed, how else could we start to unleash their immense potential for good? The possibilities seem to me to be as infinite as the microbes themselves.

The ways medicine could be revolutionised through their use are only just beginning to be imagined. If this potential were to be realised, then not only could we overcome the daily threats we face from diseases we cannot pretend to fully understand, but we could also prepare ourselves in advance for

14

the new pathogens which nature's downside seems to be constantly manufacturing to destroy us.

Microbes are starting to play a significant role in forensics. We must all be aware of the huge strides that have been made in criminal detection through the use of DNA in identifying law-breakers through the faint traces of their DNA they leave behind at the crime scene. Now work is being undertaken to try to understand if our bacterial signature can also be used to act like a super fingerprint and to trace evidence. There is already something called the Burglary Microbiome Project which hopes to be able to detect a burglar's personal bacterial signature if he or she has been in a room for ten minutes, touched some objects and left behind some of his or her bacteria.

It seems likely that our microbial signature is affected by the life we have led; whether we grew up in a city or in the countryside; whether we kept pets as a child; whether we work in a factory or on a trawler. Everything we do and experience helps to determine our bacterial signature.

Even more exciting is the prospect of using the microbiome information inside our bodies to personalise medicine, for example to predict whether we will respond to a particular therapy. At present we have a pretty haphazard way of identifying and treating illness. How much better it would be if, through the study of microbes, we were able to better understand what is actually happening inside our bodies. Instead of the current somewhat random approach to treatment based on the average care experience of a whole population, it may soon be possible to screen our individual microbiomes and determine if we have the biomarkers that can indicate our likely responses to, for example, chemotherapy. In this way doctors will be able to identify the best therapy and to prescribe tailor-made treatment. This could revolutionise medicine in many ways, not least in preventing pandemics.

These exciting possibilities lie in the near future, but already the gut microbiome is a hot topic and, not surprisingly, many commercial enterprises are jumping on the bandwagon. An interest in the diverse communities of bacteria which inhabit our intestines has resulted in our supermarket shelves and pharmacies stocking a wide array of probiotic products containing live bacteria and yeasts. Some are more effective than others and recent research has indicated that many of the commercial brands contain hardly any bacteria. But the principle has been widely accepted and acknowledged and the idea that one's general health can be improved by the addition of healthy bacteria is taking hold.

Some people have given up soap, so as to give their bacteria the best chance to do the job for them. They just use water and claim that their skin microbiome is more healthy as a result and that they do not smell. The argument is that with greatly increased bathing over recent decades there has been a marked rise in skin diseases. Now there is a growing industry providing hair, face and body washes free from preservatives and detergents. This appears counterintuitive, but it seems to work. Rastafarians and others with dreadlocks have always claimed their hair ceases to smell soon after they let it grow out. Now we are getting a glimmer of why that may be, although there is no medical evidence to support the idea yet. A lot more scientific research is needed.

AGRICULTURE

Then there is agriculture, upon which our survival depends, because without food, we die. Here a revolution is already beginning to take place.

Over the last thirty years, world food production has increased by 17 per cent, but a billion people are still hungry. So far, much of the debate, often very acrimonious, has been about whether genetically modified (GM) crops are good or bad; whether organic or non-organic farming methods are best practice; and how to make intensive agricultural production less environmentally harmful. At the heart of the problem has always been the lack of sufficient land on which to grow food and, as a result, the constant erosion of the remaining wildernesses, especially rainforests. Now we have other choices, which just might save the day.

For a civilisation, whether ancient or modern, to survive it needs to be sustainable and, in agriculture, microbes play a pivotal role. All agriculture is dependent on a symbiotic relationship with microbes: its biological processes, such as decomposition, nutrient cycling and nitrogen fixation, are regulated by microbes. Today, the extreme application of chemicals almost everywhere has deprived the soil of its rich flora of beneficial microbes and carrying on as we have in the past, by applying more, is just making things worse. The time has come to reintroduce environmentally friendly microflora into farming and, today, thanks to DNA sequencing, scientists are able to select microbes that can improve the way we grow food. A healthy, balanced ecosystem lets plants grow without the need for chemical fertilisers. Much field research is currently going into the complicated interactions of plants and microorganisms as a biological alternative to agrochemicals, as well as promoting sustainable agriculture, soil health and the biological control of insect pests, without disturbing the environment. But we need a lot more research into the incredible microbial diversity of this planet, where scientists are constantly being amazed by the multiplicity of minuscule life being found as new habitats are explored.

Many people from cultures that traditionally eat a meat-based diet have turned, and are turning to, various forms of vegetarianism, veganism and other permutations. They are motivated by a complex mix of ideological and health-based reasons: disgust with factory farming and suspicion that many modern farming practices introduce noxious substances, including antibiotics, into our food, especially meat. It is now clear that for all sorts of reasons, not least the inefficiency and waste involved in most meat production, as well as the huge production of methane by livestock, which is affecting global warming, we all need to reduce our meat consumption.[5]

Fortunately, lots of alternatives are being developed, and many of them revolve around the growing awareness of the role played by microbes.

CULTURED MEAT

There has been an explosion of research into engineering alternatives for our conventional meat-based diets – and most of that research has been based on microbes. An agriculture based on microorganisms, creating what has been called a 'post-animal bioeconomy', is developing fast. In fact, many of the leading agroindustries are already investing in it. The French dairy giant Danone was one of the first to do so, buying White Wave Foods, which produces plant-based beverage brands, in 2016. Tyson foods, one of the world's largest meat-processing companies, bought a share in Beyond Meat, a company producing plant-based meat alternatives. Cargill, another major meat producer, has invested in meat-culturers Memphis Meats. The animal agricultural industry is already reinventing itself and it seems likely that it is only a matter of time before tipping

5. https://science.sciencemag.org/content/360/6392/
987.full?ijkey=ffyeW1F0oSl6k&keytype=ref&siteid=sci

points are reached and factory farming as we know it will undergo revolutionary change. This may well not mean that we all stop eating meat, just that the meat we eat will have been produced in revolutionary new ways – and may not be based on an animal at all. 'Real' meat will continue to be available, but it will become a much more luxurious and expensive product than it is today, and much less of it will be consumed. There is much land that is only really suitable for grazing by livestock and this will continue to supply healthy grass-fed animals, but it will be done in a less intensive way. Among the exciting possibilities offered by the rewilding of much of the poorer and less productive land in Europe are the many useful by-products which will become available, such as the culled wild animals proliferating in these rich new environments. These will add yet more healthy meat which has developed naturally with no added hormones or antibiotics.

These are the next and logical steps in the agricultural revolution, which is already beginning. However, agricultural evolution is taking place at the same time. Much of it is based on the growing recognition by farmers and consumers alike of the importance of soil microbes and their potential for enhancing both soil fertility and public health. Mycorrhizae, from the Greek meaning 'root fungus', represent the most important and widespread example of plant symbiosis known. The association promotes plant growth and resistance to pathogens, so that it can literally revolutionise the yields of agricultural crops under both normal and stress conditions.

In both the medical and agricultural worlds, tissue engineering is a growing discipline and, of course, closely related to research into microbes. In agriculture, the interest is in creating meat by in vitro cultivation of animal cells, instead of by slaughtering animals. It is a new form of cellular agriculture and one of the stimuli for its growth has been the interest of animal welfare groups in its development. For instance, PETA (People for the Ethical Treatment of Animals) offered a $1 million prize to the first company to bring lab-grown chicken meat to consumers. The idea was even suggested by Winston Churchill in 1931, when he wrote: 'We shall escape the absurdity of growing a whole chicken in order to eat the breast or wing, by growing these parts separately under a suitable medium.'

Cultured beefburger patties are already being produced and the forecast is that many more varieties of food produced in this way will be on the supermarket shelves within a few years.

At the moment, cultured meat is still very expensive to produce, but with mass production the price would undoubtedly plummet. And there appear to be few ethical and religious objections to the idea. There is no animal cruelty involved, because cultured meat does not have a nervous system and therefore cannot feel pain. Initial reactions from vegetarians have been varied and will doubtless lead to some interesting ethical debates. However, Jewish rabbinical authorities seem to agree that if the original cells were taken from a kosher animal, then the cultured meat will be kosher. The same would apply to halal meat for Muslims if the original cells were halal, although this is still being discussed by Islamic scholars. Debate continues among Hindus, where most reject the whole idea of eating beef, even if it is 'meatless'. So far vegans seem to be taking the same view. But there is a strong case for it environmentally. It is certainly much more efficient and eco-friendly than conventional farming, requiring only 2 per cent of the land, and generates hardly any greenhouse gases. Compare this to the 18 per cent of global greenhouse gases generated by cattle, more than the combined effect of the world's entire trans-

---

6. Winston Churchill, 'Fifty Years Hence', *The Strand Magazine*.

port system. It has even been claimed that, in theory, once the process has been started, it could be possible to continue producing thousands of tons of meat from a 'starter pack' of only a handful of muscle cells. That would certainly represent an agricultural revolution!

The first cultured hamburger was fried on 5 August 2013.

The technology is developing fast and looks set to change everything. A Finnish company, Solar Foods, is planning to open its first factory in 2021 producing food from electricity, water and air. Renewable energy is used to electrolyse water, producing hydrogen, which feeds bacteria to produce a completely new kind of food, which does not need agriculture or aquaculture. They estimate that 20,000 times less land is required for their factories than is needed to produce the same amount of food by growing soya, the crop now covering vast areas of deforested land in Brazil. According to George Monbiot, this has the potential for feeding the world and allowing the ecosystems we have trashed to recover.[7] If our insatiable appetite for soya and palm oil were to be satisfied in this way, then not only would there be no more need to cut down the world's remaining rainforests, but they could be allowed to reclaim their lost territory – and at the same time the world could be fed and hunger banished. It is even being discussed in scientific circles how synthetic foods like these could be used to help reduce another of the curses of modern life: childhood obesity.

All done with microbes, but this is probably quite a long way away. Meanwhile, lots of other solutions, some with a venerable history, are being tried.

---

7. 'Electric food – the new sci-fi diet that could save our planet', *The Guardian*, 31 October 2018.

INSECTS

Some farmers have started diversifying into mass-producing products based on insects, which seem alien as a foodstuff to the Western world but have been happily consumed by most of the global population since time began. Commercially, this has made most sense initially as a way of substituting easily produced and cheap protein from insects to replace the dangerously destructive mass use of krill and other marine organisms in animal feed. It seems likely that with time the direct consumption of insects by humans, rather than as a concealed part of meat production, will become not only acceptable but a culinary revolution in its own right. Some supermarkets have already been trialling insects as fashionable snacks and there are cafes and restaurants which are starting to serve them.

Farming insects also addresses another of the growing concerns about conventional agriculture: cruelty. Anyone who has bothered to start looking into how most of the meat sold on supermarket shelves is actually produced is likely to have been appalled at the conditions in which cattle, pigs and chickens are sometimes kept when being factory farmed, especially in countries which have less stringent laws than those in the UK. Images of chickens spending their short lives in tiny cages before being slaughtered and cattle kept in feed lots and fed on hormones and antibiotics are what have caused many to become vegetarians. With insect farming, however, animal welfare is considered to be much better because raising large numbers in small spaces is close to their natural conditions and so they are not stressed by overcrowding. Unlike when mammals or birds are kept in tightly confined enclosures, causing stress and the risk of infection, if natural conditions are mimicked for insects, they do not need feed additives or medicines to prevent health consequences. Moreover, when it comes to killing them, this can be done humanely by reducing the temperature until they go to sleep and then, after a day or two, they die.[8]

VERTICAL FARMING

Vertical farming is another revolutionary idea which is gaining ground. This is the practice of producing food in vertically stacked layers, such as skyscrapers, warehouses or shipping containers. Controlled-environment technology manages the humidity, temperature and light. There is software that ensures all the plants get exactly the same amounts of each of these requirements and of the necessary nutrients. No herbicides or pesticides are required and everything can be located right in town, so that there are no transport costs, with all the environmental damage and 'food miles' they cause. It sounds too good to be true, but although there are certainly problems concerning energy use and associated pollution, it is likely to be another of the factors in the forthcoming agricultural revolution. And one of the best things about the idea is that it removes the pressure to find more conventional agricultural land by cutting down rainforests and draining wetlands.

Above all, these new technologies all recognise the vital role microbes will have to play in our lives in the future. Those at the cutting edge are claiming that cultivating a robust, healthy ecosystem indoors, optimised for the health of the microbiome, is about fifty times more powerful than producing it out of doors using fertilisers.

One of the most promising developments is aquaponics, a bio-system which integrates fish farming with hydroponic

8. See   https://www.sciencedirect.com/topics/agricultural-and-biological-sciences/feed-additives   and   https://www.sciencedirect.com/topics/agricultural-and-biological-sciences/insect-farming

vegetable and flower production to produce a symbiotic relationship between the plants and the fish. The nutrient-rich
waste from the fish tanks 'fertigates' the hydroponic beds,
which in turn filter the water and provide habitats for bacteria,
which augment the nutrient cycling and complete the process.
Thus the whole system is regulated – by microbes!

This developing agricultural revolution could potentially dis-
rupt the entire system. With luck it could in time solve most
of the great problems we currently face by both feeding the
world's population and restoring lost habitats at a stroke. As
the technology for producing synthetic protein makes it ever
cheaper to produce, there will be a price to pay for many traditional farming methods: livestock and fish farming will be
replaced; arable farming will be greatly reduced; and there will
be a huge fall in demand for grazing land. But this land will
then become available for rewilding, reforestation and other
habitat-restoration schemes. If soya beans are no longer needed
for cattle feed, the remaining Amazonian forests will be spared
and may even start to expand again. If a synthetic substitute
for palm oil is found, then the same may happen in South-
East Asia. Pollution from the chemicals and effluent released
by current farming methods will be massively reduced. And, of
course, modern agriculture contributes hugely to greenhouse
gas emissions, about a third of it arising as a result of what is
politely called 'enteric fermentation' or, in other words, cattle
and sheep farting and belching...

At the same time there will be a colossal drop in use of
antibiotics, 80 per cent of which are currently given to farm
animals in some countries.[9]

9. See https://www.rethinkx.com/food-and-agriculture and George Monbiot,
op. cit.

All these impending medical and agricultural revolutions are just steps in recognising how the vast microbial universe represents by far the most exciting potential for saving our threatened planet. There are sources of nutrition just waiting to be explored.

A whole new perspective on how it all works is developing – by looking inward rather than outward.

The rise and fall of the Classic Maya

The rise and fall of the great Maya civilisation is both an example and a parable. There is much to be learned from what went wrong for them. It is the story of what happens when a civilisation self-destructs through destroying its environment. Before we examine what led to the end of the Maya civilisation – and thereby what might threaten our own – it is necessary to trace its beginnings.

Let us start in the year AD 657. This was when Yuknoom the Great of Calakmul conquered Tikal after a great battle. He was one of the most successful and powerful of all Maya kings and continued to reign until well into his eighties. This was not all that unusual, some kings reaching their nineties. The moment when he achieved pre-eminent power over the central lowlands could, perhaps, be seen as the highest point reached by any of the many remarkable societies which evolved over a relatively short period of time from the small bands of nomadic hunters who had slipped between the continents a few thousand years before. It is now generally agreed that humans, after spending 100,000 years or so spreading out from Africa across Asia, finally reached the Bering Strait. One attribute the human race shares with few species other than rats and cockroaches is an insatiable urge and ability to colonise every corner of the earth's surface. It was therefore inevitable that, as soon as we could do so, we would enter the Americas, where humans had not evolved, and nor had any of the other great apes. (It is amusing to note that by far the largest population of people who deny our relationship with and evolution from apes should live there today.) The last Ice Age had locked up much of the earth's surface water in ice around the poles and sea levels had dropped. For a while a land bridge existed and so, somewhere between 15,000 and 25,000 years ago, humans entered and started to exploit the rich resources of the New World. Within a few thousand years, they had migrated all the way to its southern tip. Tierra del Fuego was occupied extraordinarily early, maybe as much as 9,000 years ago. Meanwhile, in the rest of the continent, people had begun to settle, to practise agriculture, to build towns and so to start civilisations.

The earliest people we know much about were the Olmecs, often called the 'mother culture', who originated on the coast of the Gulf of Mexico and, for some 500 years before the turn of the millennium, built temples and sculpted the massive, square-jawed figures which remain so characteristic of later Mesoamerican cultures. The Olmecs also probably developed the practice of human sacrifice, which the Maya and Aztecs took to even greater heights. They may have been the first to invent writing in the New World, although recent research suggests that the credit for this may, after all, go to the Maya, who have long been credited with it.

ART AND TIME

The art in the region continued to evolve, but it was a long time before Western eyes started to appreciate it. The British architect and topographical artist Frederick Catherwood and the American lawyer, diplomat and writer John Lloyd Stephens were the first to rediscover the Maya ruins and, through their illustrated books, they brought them to the attention of the Western world in the 1840s. Malcolm Coe, the pre-eminent scholar of the Maya, says that Maya archaeology began with them. Later, no less a figure than Arnold Toynbee was to observe, in his monumental *A Study of History*, that 'in the modelling and painting of its pottery and, above all, in its plastic portrayal of the human countenance, [its art] is not unworthy to be compared with the art of early Hellas'. He had a particularly soft spot for the Maya, whom he regarded, perhaps mistakenly, as especially cultured and pacific compared to other Mesoamerican societies. He maintained that they 'stand to the Aztecs as the early Sumerians stand to the Assyrians'. Later still, Henry Moore, after visiting Mexico City's Anthropological Museum, was influenced in his reclining figures by Maya chacmool sculptures.

A Maya chacmool figure

Reclining Figure by Henry Moore

The Maya plotted the movements of the moon and planets and their calendar was more accurate than either the Julian or Gregorian calendars of Europe. They calculated sidereal time, the time it takes the earth to complete one rotation (23 hours, 56 minutes, 4.091 seconds), to within 0.000069 of a day of the figure used by scientists today. They used the zero in their calculations centuries before we in the Old World did, and their incredibly complex Long Count calendar is still not fully understood. Its year dot is 11 August 3114 BC. This is what the Maya thought of as the date when the current world began, the founding of the current universe. It is only about a thousand years later than the rather arbitrary date of 4004 BC calculated by Archbishop Ussher in 1650 as the moment when God created the Garden of Eden. The Maya Long Count calendar has a cycle of 1,872,000 days, which is just over 5,000 years. The fifth and final cycle ended on Friday 21 December 2012, which some misinterpreted to mean the end of the world. Or perhaps Armageddon in the form of a great flood, earthquake or other cataclysm would annihilate our corrupt world, which would then start a new and more perfect world. Some might say not before time. Nothing happened – or did it, without our noticing properly, yet...?

CRUELTY AND SACRIFICE

As other descriptions of Maya life will show, their civilisation can validly be compared to ours as having reached a very high intellectual and material level, yet in many ways it was as flawed as ours by greed and cruelty. And, like so many others, it collapsed at its peak. Their bodies must have contained similar trillions of microbes to ours, even though they were totally unaware of them, not having invented microscopes with which to observe them. But they will have been influencing their lives just as much as they do ours. This was just one area of knowledge they lacked and which we have.

Obsidian, natural volcanic glass, was their steel and from it they made sharp knives and other tools and weapons. The fine blades were also ideal for slicing open the chests of sacrificial victims and ripping out their hearts or, in the case of high-ranking prisoners, beheading them. A deeply spiritual people, the priesthood was very powerful and much of the Maya literary and scientific effort recorded in their hieroglyphs was devoted to placating the gods and recording the genealogies of the priests, as well as kings, who were also high priests, and nobles. Commoners are never mentioned. Kings, being related to the gods, especially the Maize God, by far the most important as the provider of the essential food crop, were also responsible for the weather and the prosperity it could bring. Some of the competing kingdoms became immensely powerful from time to time; but unlike in the case of most other cultures, none was ever able to dominate the others for very long.

They all tended to practice primogeniture, descent being usually through the eldest son, although there were occasionally powerful queens, crowned when the royal line would otherwise die out.

The initiation rites, which were compulsory for princes when they were very young, seem to have been particularly gruesome and painful. The descriptions of them, garnered mostly from surviving carved panels and paintings rather than from texts, are sometimes strangely similar to circumcision rites still practised throughout the Old World.

The young prince would be finely dressed, wearing a head-dress and jewelled necklace. An equally brightly caparisoned noble, whose official role at court this probably was, would kneel before him and ritually 'let blood' from the child for the first time by perforating his penis, perhaps with a stingray spine or shark's teeth. A crowd of other nobles and priests, all in their best clothes, would stand around watching. The princes depicted having this done to them are usually recorded as being five or six years old.

These rituals were designed to prevent chaos and if they were neglected it was believed awful things would happen. Astonishingly elegant murals have been discovered at San Bartolo, not far from Tikal, which has been described as one of the great archaeological finds of all time and as the Sistine Chapel of the Maya, as well as at Bonampak, which show in gory detail the observances being practised by the nobles, as well as the brutal tortures inflicted on captives.

Keeping the gods happy with human sacrifices when times were bad usually involved drawing blood, lots of it, to nourish the gods and keep them satisfied, although sometimes men and later women, perhaps virgins, were thrown alive into one of the deep natural wells or *cenotes* which are found throughout the Yucatan peninsula.

It seems their reasons may have been more sophisticated than a simple desire to please the gods. Mortifying the flesh may have been thought to have an intrinsic benefit both to the practitioner and to nature at large. There are echoes here of how it happened and continues to happen in other parts of the world. In Christianity, it is all about ridding the body of evil concupiscence and sin, but among Hindus and Buddhists, and among shamans, it is used more to achieve altered states of consciousness. It may also have derived from the familiar painful rituals to mark adulthood and other special occasions, still found among many Amerindian and other indigenous peoples. The early bloodletting of all those of royal blood could well have engendered a belief in the virtue of pain. Cruelty and the infliction of pain seem to be universal human traits.

Of course, the trouble with being responsible for the weather is that when things go wrong you have to take the blame. Trying to recoup your credibility and save the day by yet more sacrifices must have worn thin as a solution when the droughts continued.

LAND AND WATER

It took the Maya about a thousand years to reach their peak. During this time they had learned how to manage one of the biologically richest places on earth, the Petén rainforest, and to become the most advanced civilisation in the Western world. Since most of the region is limestone and therefore porous, with little or no surface water, irrigation was always a problem. They built vast networks of aqueducts, canals and causeways leading to terraced and raised fields. These they fertilised with mud and water hyacinth, mulching and flooding them regularly to produce the intensive cropping necessary to support burgeoning populations.

In some areas there were *bajos* or shallow lakes, which dried up outside the rainy season. They dug irrigation canals to drain and flood and generally to manage this precious commodity. These created habitats for fish and turtles, which were an important part of the diet. This was an early and very effective form of aquaculture. Occasional seasonal water holes called *aguadas* exist, and some were expanded by the Maya into reservoirs, which held water through the dry summer months. Far and away the largest in the Maya world was one of the five at Calakmul. When full, it measured 242 by 212 metres and it is still there today, although mostly silted up now. Our friendly guard, Herminio Chan Caama, who had let us sleep on the site and visit the temples at dusk and dawn, took us to see it. This involved hacking our way through some dense undergrowth along a dried-up riverbed until we came to the edge of what he proudly told us was 'the biggest aguada in the world!' There we saw some shallow water mostly full of weeds. The river used to flow into the lake during the wet season and fill it, but Herminio said that about ten years ago the man-made canal which fed the lake from the river had collapsed and now no water got through. He had ambitious plans to restore the canal and we agreed that it would make a fine project.

The major sources of water over most of the Yucatan peninsula were and still are *cenotes*. Where these circular sinkholes did not exist, the Maya dug underground cisterns with wide plastered aprons to catch the rainwater in its season. At its height, Maya agriculture must have been a sight to behold. Where once there had been virgin forest, there were fields and orchards growing rich crops to feed the many city states which competed to dominate their part of what was never a homogeneous empire. The population density at this time has been put as high as 1000 or even 1500 per square mile, which compares with those living on all but the richest soils on earth. It was probably nearer half that, but still a heavy density for a largely rural society.

Cutting down the forest with obsidian and flint tools must have been a grindingly slow business, which in an age of chainsaws we can hardly imagine. One way they may have done it, which takes time but is effective, is simply to ring-bark the tree and leave it to die. However, the Maya did have some things going for them when it came to building and decorating their temples. Limestone is easy to cut and work with flint. When fragments are burned and mixed with water, very hard plaster can be made and this was used to face the temples and make stucco friezes. There has been much speculation as to what drives a civilisation to express its power by building ever bigger and higher structures. Frenzied building may be an indication that a civilisation is nearing the point of collapse, since many societies that have gone 'monumental' have fallen apart, often soon after creating their most massive buildings.

On the rich soils of Europe, as well as in China and elsewhere, good land management can sustain productive agriculture virtually indefinitely, but not on limestone. Agriculture was the backbone of Maya society. When the crops failed the spine of that society snapped and it started to crumble. By AD 750 the cracks in the system were beginning to show everywhere. Soil erosion was reducing the fertility of the land; the rivers had started to silt up and, as we have seen in far too many other parts of the world recently where rainforests have been cut down, water quality deteriorated and fish stocks plummeted. Droughts became more frequent, as did extreme climatic events: hurricanes, floods, delayed or failed harvests. These in turn led to war. Seeing the difficulties being faced by one population, another would have sought to take advantage of the situation. These events have been well illustrated by research done at Copan, another of the great Maya centres.

There the population appeared to have exceeded the carrying capacity of the main valley by the sixth century and they were already beginning to rely on imported food. Deforestation had removed the trees right up to the pines on the highest ridges, with the result that building timber and firewood were in short supply. As a result, there was widespread erosion and signs of malnutrition and infant mortality can be detected at burial sites. Without an understanding of the microbial role in sanitation, there must have been some serious pollution, causing epidemics and sickness.

The Maya, like so many other transient civilisations, managed their environment superbly to achieve maximum productivity from it. But they failed to understand how rapidly systems will break down if biodiversity is drastically reduced. We do not have that excuse and we have many advantages and insights they did not have, such as the key role microbes play in holding together nature's incredible variety!

Gradually, things began to fall apart. Classic Maya civilisation started to disintegrate and within a hundred years or so – in the eighth and ninth centuries AD – between 90 and 99 per cent of the population had died of disease or famine, perished in conflicts or simply migrated elsewhere. This was one of the greatest human catastrophes ever to strike a society, but there are many similarities to what is happening around the world today as countries destroy their remaining rainforest, creating desertification and forcing the local population to move.

The many catastrophic battles between warring states are well documented on stelae and the history of this period is gradually being pieced together. The Maya writing system is still far from fully understood. There were undoubtedly detailed records of much Maya history stored in the many thousands of books written by scribes on long strips of paper made from the inner bark of a fig tree. This was soaked and the fibres glued together, like Egyptian papyrus, before being treated with a smooth white finish. A thin brush was used to paint coloured hieroglyphs on the polished surface and the paper was then folded like Japanese screens.

However, almost all of these priceless documents were systematically destroyed by Spanish priests determined to eradicate the pagan beliefs they represented. Only four books escaped the annihilation. Now they are carefully preserved in museums and pored over by scholars attempting to interpret from them the richness of the Maya cosmology.

It is a tragedy that religious bigotry should have so ruthlessly expunged the huge record of Maya history, learning and ritual, as well as the accompanying literature, music, poetry, science and philosophy which this, the finest flowering of culture in the ancient Western world, undoubtedly produced. The implications are more profound than just the waste. The effect of wiping out virtually the entire recorded history of a culture is to undermine it irrevocably. Astonishingly, the Maya, the toughest and most resilient of all Mesoamerican Indians, continued to fight against first Spanish and then Mexican authority right up to the present day, but without a visible history and a recorded culture to be proud of they were always going to be considered inferior by their conquerors. For centuries no one recognised how sophisticated their civilisation had been and it was easy to overlook the conspicuous signs, such as the ruined temples, and to dismiss the living descendants as ignorant savages. How different it would have been if an active priesthood had survived, their scribes exchanging philosophical ideas and debating the meaning of life with their opposite numbers in Europe.

Disease, too, may have played a part in the Maya collapse, long before the arrival of even more catastrophic European epidemics. The Maya had considerable anatomical knowledge, which was probably a good deal more highly developed than that being practised anywhere in Europe at the same time, or until much later. However, they did tend to attribute the appearance of epidemics to sexual abuse, sin and disobedience. Syphilis may have contributed to their decline and there are indications that they suffered from many other ailments and communicable diseases as well as, eventually, malnutrition. They used a wide variety of medicinal plants, often elaborately prepared, and practised bloodletting both for ritual and medical purposes, using special lancets and knives made of flint. They also performed other surgery and extracted teeth.

Despite all these other factors, the critical factor in the collapse of Maya civilisation was climate change. At just the time when the population was reaching its height, perhaps as much as 14 million, making it one of the largest civilisations anywhere in the world except China, a series of droughts set in over a period of fifty years. These were interspersed with floods and the effect must have been deeply unsettling for the ordinary people, who would have suffered most. The capacity of the land to provide food must have been stretched to the limit, leading to famine when things went wrong. At some point, confidence and belief in their masters must have weakened. When an elite loses its power, social order is liable to collapse and anarchy ensues. Certainly, the Classic Maya were not conquered by another civilisation, which is why so many magnificent buildings remain scattered throughout the forest and are largely still unexcavated, although vast quantities of artefacts have been looted from temples and graves to feed the West's insatiable lust for 'treasure' in recent decades.

When the great droughts hit the Maya throughout the ninth century they must have caused great debates. As Mayan territory was limestone with little or no standing water over much of its extent, water was key and monumental efforts had been made to control, distribute and save it, as we have seen. But water only arrived when it rained and so it had become integral to the Maya religious culture that the bringing of rain was the responsibility of the rulers. Even when the largest reservoirs were filled to capacity, there was never enough water for more than eighteen months and life must have become fraught towards the end of a long drought.

Mankind has sought to control the weather since the beginning of time. Traditional cultures all have myths and legends, practices and rituals associated with controlling the atmosphere. From the most isolated of hunter-gatherers to the great dynasties of China, power and the weather have gone hand in hand. Emperors of China prayed for rain on behalf of their people in the most magnificent and extravagant of all their ceremonies and their 'Mandate of Heaven' depended on their being successful. Studies of rain patterns have revealed that the collapse of each successive empire, the Zhou, Tang, Yuan and Ming, all came about when the monsoon failed and it was presumed that the Mandate of Heaven had been withdrawn.

The Maya kings were rain gods and worshipped as such. Gradually, elaborate rituals had been created to demonstrate the link between the intense personal involvement of the theocracy – the king, his nobles and the priesthood, involving the production of quantities of blood, royal blood being the most precious, and the torture and sacrifice of enemies and slaves, as well as children and pretty girls – and the satisfaction of the gods, who would then send the rains. When they didn't come, there must have been those who questioned the system. This usually took the form of suggesting that the royalty were not holy enough, resulting in them trying harder, sacrificing many more people in ever more brutal ways and eventually, if that didn't work, being overthrown and replaced by another dynasty. But these were very sophisticated people, who had invented some of the most elaborate mathematics and astronomy anywhere in the world at that time, as I have described earlier.

There must have been some whose questioning went deeper. After all, the Maya cosmology had been designed by man and so it could be challenged by man. As the rivers began to dry up and the wind blew tumbleweed across dustbowls where once there had been lush forest, there would surely have been some who questioned if there might be other causes for the drought than the anger of the gods. Some may even have raised the possibility that the drought followed deforestation and that it was their own actions which had caused the climate change. And, just as today, they would have been mocked and derided by those with vested interests in not believing that this could possibly be the case. Then it was because only the gods had the power to bring rain, and those who questioned that were blaspheming and quite likely to have their heads cut off. Today the naysayers are either in the pocket of industries which thrive on the destruction of the environment and the extraction of raw materials like timber, as well as fossil fuels, or they simply don't want to, or are afraid to, believe the inescapable evidence that we are rapidly pushing our planet to the verge of a cataclysm.

PEOPLE AND THE FOREST

Surprisingly, the drier lands to the north, away from the rainforest and on the barren limestone of the Yucatan, continued to support Maya communities for much longer than in the south. They were still there 500 years later when the conquistadores arrived and once again 90 per cent of the population were wiped out, this time by imported diseases to which they had no immunity. Those who did not die of disease were subjected to some of the most brutal, gratuitous cruelty the world has ever seen. From the very first moment of Christopher Columbus's arrival on the island of Hispaniola a sort of madness to torture and destroy the people they encountered seems to have seized the conquerors. Columbus was forbidden by Queen Isabella to make slaves of the 'Indians' and his clear instructions were that they should not be treated as chattels or goods. The Dominican priest Bartolomé de Las Casas, who campaigned ceaselessly on behalf of the Indians, wrote volumes about what was happening and begged that they be treated better, describes that first encounter and the many that followed. He says that Columbus and his crew were all graciously treated with the greatest kindness 'and looked after as if they had been back home and were all part of the same family'. Anyone who has been lucky enough to receive hospitality from traditional Amerindian people will immediately recognise this behaviour. Yet in no time at all the chief's wife had been raped and many Indians hacked to death, while others fled to the hills. Time and again, as successive Spanish ships arrived at new islands and, in due course, the mainland, Las Casas tells of the treachery of the conquistadores as they tricked the trusting Indians, who greeted them with gifts, into becoming their prisoners. Once bound, they were subjected to as horrific and gratuitous torture as has ever been perpetrated anywhere. Ostensibly, this was to make them reveal where there was gold, which the Indians soon recognised was the Spaniards' true god, rather than the Catholic one to which the priests urged them, often with the greatest ease, to bow down. A vicious insanity seems to have overcome the original conquistadores and the Spanish settlers who followed them. Tens of thousands of Indian men, women and children were flogged, burned alive and hacked to pieces over the first few decades of conquest. The descriptions in Las Casas's books, especially *A Short Account of the Destruction of the Indies*, first published in 1542, are sickening and what makes it all even less comprehensible is that this good priest and a few others were constantly revealing to the authorities what was happening, confronting the perpetrators and begging them to stop. But 'anaesthetized to human suffering by their own greed and ambition', the Europeans who had 'discovered' these highly civilised peoples continued to kill and enslave what Las Casas calls 'those vast and marvellous kingdoms' until few were left alive.

The region which had been most abandoned in the ninth century was the heart of what had been the Maya empire, where Tikal and Calakmul had been the two most powerful of all Maya cities. This happens also to be the heart of the Petén. Here, due to the virtual absence of man, the rainforest was able to recover. By the nineteenth century, after a thousand years of tranquillity and solitude, it was almost completely restored. The environment was once again intact and teeming with life. There were hardly any people to hunt or chop down the trees. It is one of the rare examples anywhere in the world of this happening, since elsewhere population pressure has usually meant that new migrations of people, sometimes from far away, have moved into the empty space. But it didn't happen here. It is a good example of the Gaia hypothesis, which demonstrates how well nature will manage in the absence of man. It was my friend and one-time neighbour in Cornwall, James Lovelock, who originated the Gaia hypothesis and first described how the world works as a whole, self-regulating the climate so that it is conducive to life and therefore habitable. This is done not for the benefit of man, as successive religions have chosen to aver, but rather in spite of our destructive tendencies. We are the product of evolution and subject to its laws. If we flout them, we, like so many other species who simply failed to meet evolution's ruthless rules, will pay the price, which is extinction. The reason our species has survived and prospered for so long is because we have always been able to move on, take over more land and in due course, unless it was exceptionally rich land, overexploit it in its turn. What makes our progression towards extinction so unforgivable is that, unlike all those who have succumbed before, we can see clearly what is happening and yet we choose to ignore or deny it.

Part of the answer to why the Petén recovered was because no one really wanted the land. The Spanish saw no living cities to loot for gold, the soil was not suitable for cultivation without a lot of hard work and the remaining Maya were an awkward lot, who have never really been conquered. The last Maya city, Flores on Lake Itza in today's Guatemala, and which the Maya called Noh Petén, did not fall until 1697. In recent centuries the relatively small remnants of Maya society have resolutely fought against domination. At the end of the nineteenth century, during the savage Caste War, rebel Maya in the Yucatan even petitioned Queen Victoria to become part of the British Empire! Instead, the border between Mexico and what was then British Honduras (today Belize) was established by the Mariscal–Spencer Treaty in 1889. Even the Zapatista National Liberation Army, which has been in rebellion against the Mexican government since 1994, is today largely composed of various Maya groups.

And now, predictably, the forest is under threat again. Over 800 square kilometres of the Maya Tropical Forest is being destroyed each year. Successive droughts are making the situation worse. A farmer we spent time with in Guatemala told me that in his father's day the rains always used to arrive without fail on the fifth of May each year and continue for eight months. Today there is often no rain until August or September and it only lasts for a couple of months. He was in no doubt that the climate was changing. 'You have only to look at the clouds,' he said.

DAILY LIFE UNDER THE MAYA

The Maya empire's growth was based on their intimate under-standing of their environment and their meticulous use of it. The myriad ways they exploited the extreme biological rich-ness of their world demonstrate the huge variety of valuable and nutritious life within a rainforest and justify our desire to protect it.

Most academics agree that there was a degree of cruelty in the running of Maya society and that is hard to argue with, since terri-ble scenes are portrayed in so much of Maya art. However, it cannot have been like that all the time. I believe that for most people most of the time life would have been very different and these represen-tations of horror were made at significant times of stress. The equiv-alent would be if history branded the whole of European society as cruel because surviving art portrayed the crucifixion story and the torture of martyrs.

For the vast majority of Maya, daily life would have revolved around many of the same activities as it does today. The four bases of their diet were the same: maize, beans, chilli peppers and squash. Of these, maize was far and away the most impor-tant. It is a highly productive crop but, in its untreated form, it lacks or locks up the amino acids and niacin which are essential to human health. A diet dependent on untreated maize results in malnutrition and pellagra, a nasty disease which causes diar-rhoea, a swollen tongue, severe rashes all over the skin and dementia. Somewhere in the dim and distant past, perhaps dur-ing the time of the Olmecs 3,000 years ago, someone found that, after drying the ripe kernels and removing them from the cob, boiling the maize in water and white lime for an hour, throwing away the water (which is by then poisonous) and grinding the resulting mash enhances the amino acids and frees up the niacin. This is, of course, a biological process, triggered by microbes. It has been suggested that without the invention of this special technique for preparing maize none of the American civilisations would have been able to develop. The unleavened dough is then either made into *tamales*, when it is wrapped in a leaf, or *tortillas*, little flat cakes, which are baked on a griddle or on the ashes of the fire. When the Spanish brought maize to Europe, they failed to bring the technology for cooking it with lime (nixtamalisation), which makes it suitable to use as 'bread'. With their more powerful mills, they thought it unnecessary to soak the kernels before grinding them. As a result, there were epidemics of pellagra and, except for 'sweetcorn', in Europe the bulk of maize was and still is fed to animals.

Maize, among the most productive of all food plants and the staple diet of Central America, is grown in quite different ways in the highlands and lowlands. In the highlands, where the soil can be rich and deep, many different kinds of maize are grown and they support dense populations. On the poor forest soils of the lowlands, where the greatest Maya civilisations developed, a completely different technique has to be used. Throughout history and still today, this has been the clearing of plots, or *milpas*, by slash-and-burn shifting cultivation, a practice which works extremely well when done properly. It has been described as one of the most successful human inventions ever. A farmer will clear a hectare or so of forest, leaving some trees for shade. After burning the felled trees and brushwood, he plants the maize seed, ideally just before the rains come, and among it the secondary crops of beans, chilli peppers and squash. After a couple of years, he will abandon the *milpa* and leave it fallow for between four and twenty years, depending on how poor the soil is and how long the forest will take to recover. As we have seen, the Classic Maya created for a time a quite different system of intensive farming, but the land was eventually exhausted, which undoubtedly contributed to and may well have been the cause of the sudden collapse.

Beans, both red and black, were an important source of protein in the Maya diet. They were often mashed and wrapped in tamales and tortillas, as they still are today, strongly spiced with chillies. Squashes, pumpkins and calabashes played a significant role, too, as some of the staples. Calabash squashes were used as containers, while the flesh of vegetable squashes was eaten raw and the seeds dried and roasted to be eaten as a snack. Rice, wheat, chicken and pigs, which are today integral to the modern Maya diet, were introduced by the Spanish and unknown to the ancients, but many other fruits and vegetables were cultivated, such as avocado, tomatoes, sweet potatoes, papaya, guava and yucca or manioc. Cotton was grown for making into textiles, which the Maya exported to other regions. Vanilla was cultivated widely. There are thousands of species of orchid to be found in the Maya rainforest, but only one of them has edible fruit, the vanilla orchid, which is now cultivated widely in the Philippines, Tahiti and elsewhere in the South Seas. It all originates from the plants gathered and propagated by the Maya since earliest times. Starting in the sixteenth century, Spanish trading ships called Manila galleons used to sail across the Pacific to the Philippines from Acapulco, taking New World silver and a few other products and trading them for Chinese porcelain and silk, and spices from the Moluccas. This was how vanilla plants reached the Far East. Today most vanilla is produced in Indonesia and Madagascar. Vanilla is perhaps the best known flavour in the world. This is because of its complex, subtle taste. It is now known that microorganisms are present during the curing stages and research is beginning into the role they play in generating variations in the flavour. This is a whole new field of science and it is revolutionising our understanding of food.

THE COCOA BEAN

Many wild fruit and other plants were also gathered from the forest and used as flavourings, burned as incense or consumed by the Maya, but the most important fruit of all to them was the cocoa bean. Most people today have little or no idea where chocolate came from. It was the luxury drink of the Maya nobles for more than a thousand years and the cocoa bean, from which it was made, was the prevailing currency. The tree on which the bean grows was given a delightful and appropriate name by Linnaeus: *Theobroma cacao*, meaning 'food of the gods', and so indeed it was perceived. There are many contemporary pictures and carvings of Maya gods carrying cocoa beans, and venerable beans have been found in a ruler's tomb. The Aztecs, who subsequently occupied most of the Maya territory and demanded tribute in cocoa beans, believed that their god Quetzalcoatl descended from heaven on a beam of the Morning Star (Venus) to bring the cacao tree to earth and teach the people how to make chocolate from it. They adopted the Maya name for it, *xocoatl*, which means 'bitter water' and they only consumed it as a drink, never in the solid form we know today.

It has been suggested that the drink was first produced as an alcoholic beverage made from fermenting the pulp around the seeds and that it was only discovered later that a better drink was made from the seeds themselves.

Whatever the truth of its origins, it was recorded that Montezuma drank chocolate from golden goblets and was said to consume fifty draughts a day. Unsweetened, rather oily and often drunk cold, it would seem not to suit modern tastes, but it was highly prized and for many centuries ordinary mortals were not allowed to drink it. It was also used extensively as an ingredient in various other dishes and drinks, often mixed with chilli pepper to give it 'burn', and as a flavouring in stews. Whipping it to create foam made it much more desirable.

Today, chocolate is very big business indeed, but for a long time it brought little or no benefit to its original inventors. After the conquest, the Spanish took control of the Pacific coastal plains where the cacao plantations were and they brought the drink to Europe for the first time. By the seventeenth century, now sweetened with sugar, it became a craze and the colonists started to create plantations worked by slaves in other parts of central and South America and in the Caribbean. Adding milk, thus making the drink so familiar today, was invented by Sir Hans Sloane when he was in Jamaica in 1687. Previously, it had always been drunk mixed with water. By the nineteenth century, Cadburys were selling his recipe as Sloane's drinking chocolate. Sloane's extensive collections were to become the founding core of the British Museum and the Natural History Museum.

In the nineteenth century, the beans were being grown in West Africa for the first time and today well over half the world's production comes from there, with less than 2 per cent coming from Mexico, where the cocoa bean originated. The Maya seemed to have been left behind by this huge industry, until a company called Green & Black started producing dark, Fairtrade Maya Gold chocolate in 1994. This is made, in part, from cocoa beans grown by the Maya, who share in the profits and continue to use small plots on their own land in the forest, rather than growing them on large plantations with heavy use of fertilisers and pesticides.

What neither the Maya nor later chocolate makers understood until quite recently was that microbial activity is essential both for converting the bitter inedible seeds into chocolate and for delivering the flavour. The quality of chocolate is influenced by bacterial diversity during fermentation and drying, a crucial factor in creating the taste. When harvested, the cacao pod is split open, exposing the sticky pulp which surrounds the beans. Seven days of microbial fermentation then follow, during which temperatures as high as 50C are reached and microbial products such as lactic acid and ethanol kill the beans and trigger the special flavours we all enjoy. Up until now, this fermentation has been a spontaneous process, relying on the natural microbiota at cocoa farms. But now scientists are looking at ways of creating new starter cultures, which should produce more reliable and better-quality forms of chocolate and who knows what new and exciting flavours.

ALCOHOL AND HALLUCINOGENS

The Maya did have alcoholic drinks from early times. Mostly, this was maize beer or *chicha*, where fermentation is achieved by spitting (normally by women) into the brew. Amerindian people throughout the Americas had discovered that naturally occurring microbial enzymes in saliva catalyse the breakdown of starch in the maize into maltose. This process of chewing grains or other starches was used in the production of alcoholic beverages in other pre-modern cultures around the world, including, for example, sake in Japan.

Variations of the practice are to be found throughout the Amazon Basin, where chicha is usually made from cassava. Traditionally, the women chew the washed and peeled cassava and spit the juice into a bowl. Cassava root is very starchy, and therefore the enzymes in the preparer's saliva rapidly convert the starch to simple sugar, which is further converted by wild yeast or bacteria into alcohol. The juice is then left in the bowl for a few hours to ferment; I have often been offered it by Indian groups. It tastes a bit like sour skimmed milk.

Among some of the Maya only old people over sixty were allowed to drink alcohol. Young people caught doing so were put to death. In the early days, as recorded graphically on ceramics and hieroglyphic inscriptions, it was mostly used for ceremonial purposes. To speed up the effect, the alcohol seems often to have been administered by a bone enema. Suppositories are well known to deliver medicine more effectively and swiftly than when taken orally and apparently the same goes for alcohol and hallucinogens administered rectally, which are then immediately absorbed by the body and the desired effect hastened.

On arrival in the Americas, the conquistadores were surprised to find the use of enemas being widely practised and it even led some of them to conclude that the Indians were prone to committing sodomy. Today, the paraphernalia involved – bone tubes and leather or rubber bags – is being recognised in Maya tombs, having previously been something of a mystery. This has been the result of the discovery of some graphic illustrations of the process on ceramics, where old men are shown having the treatment administered by young girls.

This technique is, it seems, highly efficient at inducing hallucinogenic visions. It is quick and safer, too, since there is less risk of infection, and very different from the dangers modern drug addicts face from injections; it is something the Maya may have been better at than us, without necessarily knowing about germs.

The administration of medicines through the backside rather than the mouth is not very common in the English-speaking world, but it has been familiar in many societies throughout history and is still much used today. The nearest some of us in Europe have come to it is the embarrassment of finding, when collecting a prescription in a French ski resort, that the large pills are suppositories and not meant to be swallowed. Some alternative medical treatments for cancer involve the use of coffee enemas, and colonic irrigation has become a fashionable and popular part of the treatment in many health spas.

Other frescoes show part of a ritual where the recipient of the enema is lying on his side while an elaborately dressed woman wields the 'clyster' or enema syringe and a speech scroll comes out of his mouth. The liquid, probably either mead made from honey or *pulque*, the fermented juice of a cactus, was fortified with datura or jimsonweed, a plant found all over the world, and is seen being prepared in large pots while vultures fly overhead. Datura has been used as a mystical sacrament throughout both the Old and New Worlds since ancient times. It is said to elicit visions of serpents, which were an essential part of the Maya cosmology.

A Maya figurine showing self-administering of an hallucinogenic enema

Another advantage of this method of taking hallucinogens is that it avoids the nausea and other unpleasant side effects which usually accompany administration. I have watched Yanomami shamans having hallucinogenic snuff blown up their noses through long bamboo tubes. This is a quick and highly effective method of entering the spirit world and they do it quite regularly; but it can be an unpleasant and painful experience. Mucous and vomit stream from the nose and mouth and the passage between the worlds is pretty revolting, hardly encouraging abuse by the uninitiated, although many have tried. In the case of the Maya, it has been suggested that the grotesque masks worn by the priesthood may have represented a cosmic howl, what one modern practitioner described as the gut-wrenching expression of one experiencing hallucinatory ecstasy. The Yanomami shamans I observed appeared to have total mastery over their actions. The snuff they inhaled was an extremely potent hallucinogen of a type that, even taken under controlled medical conditions in the developed world, might well be frighteningly disorientating. Yet the Yanomami shamans never seemed to experience any prolonged discomfort. I also noted that regular use of the drug had apparently not damaged their health in the slightest. Even the older shamans, who had been taking hallucinogens perhaps once a week all their adult lives, seemed quite as fit, strong and intellectually alert as their fellows.

The indiscriminate use of drugs is a serious social problem in the modern industrial world. The Amerindian approach to hallucinogens, however, as I saw with the Yanomami and other tribes, and as much research has revealed, is a completely different matter. By providing a philosophical framework within which the experience can be understood and assimilated, and controlled conditions for the drugs' use, many practitioners seem to have avoided the problems of abuse. In fact, this is an area where the Amerindians are the experts and we are the novices. The psychological experiments which the shamans are constantly making in the most serious way are directly equivalent to, and arguably well in advance of, our own research. The main social function of the shamans is curing the sick. Taking a supernatural view of the causes of illness, while at the same time using their very sophisticated and ancient understanding of plants and their properties, especially when altered by microbes, to help them enter the spirit world, gives them great knowledge and power. The Yanomami naturally turn to those with transcendental powers to cure them.

This leads one to look at the relative effectiveness of Amerindian and Western medicine. Certainly the fact that the shamans consider most diseases to be of supernatural origin does not mean that as healers they are ineffective. A growing body of biomedical research indicates the importance of the patient's mental state in treating even such intractable conditions as cancer. It seems that the faith a patient has in the regimen he or she is subjected to and the relationship they have with their doctor can often be as important as the medication they receive. And, as I have described above, we are only now beginning to appreciate the part played by microbes in our health – and, of course, in the creation of hallucinogens, which are mostly the result of fungal and microbial actions. For instance lysergic acid, a precursor of LSD which is used in many medical procedures, including treatment for dementia, is made from a bacterial fungus. Seen in this light, the shaman's ability to explain a disease and to act out its cure symbolically is very effective medicine indeed. By releasing the patient's stress and anxieties, he provides the climate in which the body's own healing mechanisms can function at their best, and so greatly increases the chances of a successful cure.

Throughout the New World, the Amerindians smoked strong wild tobacco and other herbs both for pleasure and as hallucinogenic methods of divination. Hallucinogens were taken for ritual purposes, on the whole carefully and sensibly, using the drugs as a vehicle to expand the minds of highly trained practitioners who knew exactly what they were doing, never for the instant gratification demanded by so much of today's society. However, among the Maya, the use of enemas may have led to the abuses which seem to have become common when the civilisation began to fall apart. Having alcohol and narcotics administered in this fashion was, apparently, a great way to get drunk or high quickly and it seems likely that the ritual use of hallucinogens to enter the spirit world may have become degraded towards the end, with spoilt and bored nobles using drugs for recreational purposes.

MAYA MEAT

Dog meat provided an important source of protein. The Maya had several breeds of dog, the ancestors of which it is believed they brought with them across the Bering Straits from Asia. Some were kept as pets, some for hunting and some were bred as an important meat supply. These, which did not bark, were castrated, fed on maize and six or eight tortillas a day, and then killed and eaten when they were a year old, being either stewed or roasted.[1]

Deer were probably the most significant source of protein and there is evidence that they were domesticated and kept inside houses, as well as being farmed. There is also a record of the conquistadores coming upon a sacred herd of deer, which had no fear of man and so were easily slaughtered. Most of the three types of deer found in the Petén, the white-tailed and two varieties of brocket deer, were hunted for their hides and their meat, but they can become quite tame when left in peace. In the early dawn, just after sunrise, as we climbed down to the forest floor below Temple I at Calakmul, we came upon

1. See https://elvalleinformation.wordpress.com/spanish-war-dogs/

a small herd of half a dozen brocket deer. Seen through the morning mist, they passed like wraiths through the trees below the temple steps, but they did not run away when they saw us. One stopped and stared intently at us, then stamped his feet and pronked for a few paces, before continuing to graze. Maybe they were the distant descendants of a sacred herd.

Tapir were hunted for their plentiful meat and their tough hides were used to make shields and armour. Jaguars played an important part in Maya rituals and their skins were eagerly sought for ceremonial purposes. In one royal tomb the decomposed bodies of fifteen sacrificed jaguars were found. Both spider and howler monkeys were a reliable food source. Turkey was another major food. One of the six subspecies of the American wild turkey from which all modern farmed turkeys descend, *Meleagris gallopavo mexicana*, came from the Maya forest. This particular species of wild turkey has a white-tipped tail, unlike their native northern relations, and that characteristic is now found in most domestic turkeys throughout the world today. Both this and the spectacularly lovely ocellated turkey, *Meleagris ocellata*, which is only found in the Yucatan Peninsula and nearby parts of Guatemala and Belize, were domesticated by the Maya for their meat, eggs and plumage, which was much used in ceremonies for necklaces, headdresses and for feathering arrows. The Maya also traded these with their neighbours. The Maya may have been the first people to domesticate turkeys, but by the time the Spanish arrived they were to be found being reared by people far to the north.

The Maya kept bees in hollow logs as well as gathering wild honey. In those days, the only bees in the Americas were small stingless bees which produced rich, dark, delicious honey, highly prized by the Maya. Since the 1950s, when some hives of particularly venomous European honeybees were brought from Africa to Brazil and almost immediately escaped, the indigenous species have been rapidly displaced by these so-called 'killer bees'. They are not as dangerous as some sensational books and films have made out, but they have caused havoc with traditional beekeeping practices. Many modern Maya do still keep bees, but now they have to treat them with respect, since the stings are very painful. There may be less wild honey, too. Once again, our growing understanding of microbes has a very significant potential role since bees, too, carry a burden of microbes in their gut, although many fewer than us. As I explained above, we have tens of trillions of microbial cells in our intestines, made up of perhaps a thousand different species, and research into our microbiome is incredibly complicated; honeybees are thought to have only about one billion, composed of less than a dozen species. As with us, these bacteria play an important part in fending off parasites and creating nutrients, but as there are so many fewer of them, scientists should be able to analyse their effect much sooner and that may help them restore the balance for the bees.

The Maya always were and still are hunters. Bows and arrows did not reach them until after the end of the Classic period and so in those days they would have used spears, darts and nets. To make the projectile go faster and truer, they used an *atlatl*, a throwing stick, which helped the spear reach 100 metres at as much as sixty miles per hour. Atlatls have been around since prehistoric times and are found all over the world, from Australia to Europe. The earliest recorded one is from a European site dated at 17,500 years ago. Today the principle can be seen in the plastic toys used for throwing tennis balls a long way for dogs to fetch.

Traces of almost everything that could be hunted and eaten by the Maya have been found by archaeologists. Among the larger mammals consumed, apart from deer and tapir, there were peccaries, the fierce American pig, unrelated counterpart of the wild boar. A wide variety of smaller mammals ranged from the delicious and highly prized pacas and agoutis to rabbits, armadillos, monkeys (both howler and spider), coatis, coyotes, badgers and squirrels. Iguanas were and still are an important part of the local diet. The meat of the green iguana, known in Belize as bamboo chicken, is said to be preferable to that of the black and today there are large iguana farms in parts of Central America. It is reputed to be an aphrodisiac. When travelling with the Choco Indians in Panama many years ago, I regularly ate iguana eggs, gathered on the riverbank. They were particularly good scrambled.

They must have eaten many insects, too. Mexico is known as the entomophagous (insect-eating) capital of the world, with some 230 species of insect recorded as being eaten. Insect eating has been rare in the modern world, but it used to form an important part of the human diet and is still widely practised throughout the tropics. I once lived for a while with the Yanomami Indians of northern Brazil. Quite a large part of their diet consisted of various sorts of insects. One of their staples was termites, which abounded in their forests. It has been estimated that there is over a ton of termites in every hectare of land, a far greater biomass of nutrition than could ever be replaced with cattle. I often came upon parties of women at work on a termite nest, which they had dislodged from the side of a tree. The big, spherical structure had been cut in half and the two halves perched open side down on the end of sticks, like umbrellas. One woman tapped each half in turn with a machete. The small black termites and the much bigger white grubs fell out onto leaves placed to catch them. The grubs were particularly rich in food value, having, weight for weight, three times as many calories and as much protein as prime beef. Once separated from the adult termites, which were discarded, the grubs were either roasted in a leaf parcel or pounded in a wooden pestle and mortar until they formed a paste. The roasted ones tasted rather like soggy breakfast cereal, while the pâté reminded me of Shippam's fish paste. It pleased me to think that a few spoonfuls of the pâté provided me with the nutritional equivalent of a pound or more of prime-cut steak.

What is even more amazing, and highly relevant for my thesis in this book, is that the termite gut has one of the highest microbe densities on earth.[2] New research into the symbiotic relationship between termites and their microbes is leading to a better understanding of how this enables them to limit methane production and break down material in their gut, such as cellulose. This research may by extension help us understand how to reduce methane production by cows and improve the production of biofuels. All this from a tiny creature no more than half an inch long. No wonder people find the infinity of small things as difficult to comprehend as the infinity of space, but it is important that we try.

The Yanomami also collected various sorts of caterpillars, being careful to avoid the ones which had dangerous stinging hairs, but gathering the edible ones which collected in clusters on leaves. These were usually wrapped in leaves and baked until they were crisp and had the consistency of whitebait. I had no problem eating those with gusto, but on one occasion I saw some green and brown specimens being prepared. They were only lightly cooked and the leathery body casings concealed soft centres that burst in the mouth on chewing. I did find it difficult to simulate delight as those were pressed on me. By contrast with termites and most other herbivores, caterpillars have virtually no gut microbes.[3] Scientists find this dis-

2. Jared Leadbetter, 'The termite gut and its symbiotic microbes', https://www.ibiology.org/ecology/termite-gut/
3. See https://www.sciencedaily.com/releases/2017/08/170822123847.htm

turbing, as they are just beginning to realise how important microbes are to almost all other creatures. What does it mean? The jury is out.

Much pleasanter were the young bee grubs we would eat first on acquiring a honeycomb of wild bees. These were considered a particular delicacy by the Yanomami. Their flavour was strong and quite distinctive; it reminded me of fresh bamboo shoots with a trace of sweet-and-sour sauce.

Tribal people will have much to teach us as we begin to experiment with eating insects, and so will research into the Mayan diet.

Much of daily life under the Maya must have been pleasant enough and no different from that of early societies all over the world; their resource base was rich and they exploited it to the full, and so built their empire. No one farms today like the Maya did and no one is quite sure what it would have really looked like, but we can imagine. The agricultural industry required to support the huge effort which went into building these great cities, the equal of anything anywhere in the ancient world, must have been immense.

DANCE AND KNOWLEDGE

Rows of small, graded turtle shells arranged just like a modern marimba have been found in tombs – probably musical instruments played like a xylophone. No one knows what the dancing would have been like, but we can speculate from the number of percussion instruments used by the Maya that rhythm must have played a significant part. I have often seen tribal dancing by Amazonian Indian tribes, where circles of men and women, usually separately, will stamp and chant in rhythmic harmony. These are sometimes not dissimilar to the dances of many North American Indian tribes and it is fair to assume that it is a style typical of the pre-Colombian world. While the dancing in the sophisticated, and later degenerate, royal courts of the Maya would certainly have evolved into all sorts of forms we can hardly imagine, it seems likely that some echo of the original rhythm and form would have continued.

In Maya society, dwarfs seem to have been regarded with some special awe, since they were associated with caves and the underworld. So, too, were centipedes, which are often depicted in Maya art as fantastic, gigantic, skeletal snakes with prominent teeth. They were the intermediaries of the underworld, associated with death and decomposition which, like the monsters in Hieronymus Bosch's paintings, drag their victims down to hell. It is not surprising to me that centipedes were regarded with such dread by the Maya. When I was living in the Borneo rainforest, there was only one animal we were afraid of – and that was the centipede. They would hide in our clothing and the pain from their bite was said to be worse than childbirth, as was confirmed by the wife of our doctor, who was unlucky enough to be bitten by one secreted in a tea towel. Centipedes are another underexplored group of rainforest creatures which demand further research. Perhaps because they live and hunt in decaying organic matter teeming with bacteria, fungi and other harmful organisms, they have been found to have medicinal properties. The active ingredients can be extracted from them and could prove valuable.

All seemed well with a very prosperous world, but there were hidden horrors, some of which are only now coming to light as new frescoes are unearthed and more stelae translated. The savagery of the Spanish conquistadores and the ruthless destruction of so much priceless and magnificent Maya art, as well as almost all their written records, led for a time to a school of thought which tended to see the Maya as a gentle people overrun and abused by brutal Western invaders. This has been true of so many wonderful societies which have been sub-jected to extreme violence and brutality that it is easy to believe unquestioningly. Certainly, they suffered as badly as any pre-Colombian people from the conquest, even though they held out longer than most. But their peculiar obsessions with blood and sacrifice have to be acknowledged.

The Maya and other ancient civilisations exploited their environments in extraordinarily sophisticated ways, learning by trial and error what was edible, what poisonous, etc. As we are now beginning to realise, the immense diversity of life on earth is dependent on the trillions of microbes that inhabit every corner of all life, including our bodies. They didn't know that; we do, which gives us a huge advantage – one we should be grasping.

The immense pharmaceutical knowledge of traditional soci-eties all over the world, derived from aeons of trial and error, is one of mankind's greatest treasures – and it is vanishing fast, as these societies are 'Westernised' and their wisdom despised and lost. It has been suggested that up to three quarters of the com-pounds used in the global pharmacopeia are based on plants used by indigenous people. Combining as much of this knowl-edge as can be rescued with the new cutting-edge research into the role of microbes would bring about a much greater under-standing of how life on earth really works.

What can we learn from the collapse of the Maya and other comparable civilisations? If we are to avoid the same fate, we are going to have to take action urgently. Understanding how our ecosystems work is key and that will require much greater expenditure on research.

Part 2

THE RED HORSE: WAR

The problem: Conflict

There have been few extended periods throughout history when mankind has lived at peace. Even though we have managed to avoid an international Third World War for the last 75 years, nearly as many people have died in civil wars and mass killings in the second half of the twentieth century as in the first: 35 million in China's Cultural Revolution alone, 4 million in Cambodia, 3.8 million in Vietnam – then there are Syria, Yemen, Afghanistan, Bosnia, Iraq, Lebanon, Congo, Rwanda and the list goes on. With nuclear weapons beginning to proliferate again there is, of course, still a danger of a catastrophe eliminating all life except microbes, when our planet will have to start the whole long process of evolution all over again; but in the meantime there will still be wars and we should consider how to prevent them.

Although religious fanaticism, conflicting political ideologies and simple territorial conquest are the manifest justifications for war, the underlying reasons are often to be found in poverty as a result of overpopulation. Prosperous, stable countries tend not to fight each other, although they often get sucked in to protect their interests in oil and other mineral wealth, or to fight proxy wars to maintain their influence, or sometimes to distract their citizens from problems at home. Poverty and lack of education, especially in places where the population is growing rapidly and traditional values are being undermined and threatened, trigger unrest and lead to escalating discord.

The solution: Free electricity

Radical action will be needed, but I believe we have a solution within our grasp, which could bring about a smaller, better world relatively painlessly: free electricity for all. This is not as far fetched as it sounds at first and is entirely possible. One hundred per cent renewable energy systems are now very much on the international agenda. They are all getting dramatically cheaper: solar, wind, biomass, geothermal, hydropower and biofuels. Nuclear power is another story, its Achilles heel being how to dispose of dangerous material which still has a half-life of many thousands of years. This problem also applies to storage batteries, which are improving by leaps and bounds and will soon be affordable to all, especially when the potential for all electric vehicles to be linked to the grid is fulfilled.

The idea of free electricity for all was first propounded by the eccentric genius Nikola Tesla in the late nineteenth century. He believed that, in addition to sending messages, telephone calls and images around the world, radio and microwaves could also be used to transmit millions of volts through the air. He was far ahead of his time and many of his inventive dreams are only now being consummated and acknowledged, not least by the decision of Elon Musk to call the electric car he manufactures a Tesla. He might have called it an Edison, in honour of the acknowledged inventor of the light bulb and of DC (direct current). But it was Nikola Tesla who invented AC (alternating current). These two giants of developing electric power fought viciously to discredit the other – Thomas Edison even electrocuted an elephant to show how dangerous AC was. There is a horrible, grainy film on YouTube showing this cruel and unnecessary stunt. Tesla won that battle and power lines all over the world have operated on AC ever since, although DC is making a comeback now, not least because DC can be stored in batteries, which cannot be done with AC. Ironically, it is Musk who has led the way with battery technology.

In fact, renewable electricity already has the potential capacity to generate enough energy for the entire world. One extreme example of how this might be achieved was propounded by a Berkeley, California professor, Mehran Moalem, in 2016. He suggested that covering 1.2 per cent of the Sahara Desert (43,000 square miles) with solar panels would be sufficient to meet all of Europe's energy needs.[1] There is no way, he claimed, that coal, oil, wind, geothermal or nuclear could compete with this. He estimated the cost of the project to be about 5 trillion dollars, less than the bailout cost of banks by Obama in the 2008 recession, or about a quarter of the US national debt, or 10 per cent of the world's annual GDP. A lot of money, but still small relative to other spending. Moalem also claims that there is no future in other energy systems, maintaining that in twenty to thirty years solar will have replaced all other forms of generation. Although political instability and many other constraints make such a dream difficult to deliver, one project in Tunisia is already under way, with

1. See   https://www.forbes.com/sites/quora/2016/09/22/we-could-power-the-entire-world-by-harnessing-solar-energy-from-1-of-the-sahara/

cables being laid to Europe under the Mediterranean Sea and with the potential to deliver 4.5 GW of electricity.[2] I prefer the idea of every available roof being used to generate electricity, as that gives much better security and independence. I was recently encouraged to believe that this revolution is already happening quietly, even in the remotest places, when I was on a river in the interior of Myanmar. A thatched hut in the middle of nowhere had a single small solar panel above it. I asked what it was for and was told, 'For my mobile phone, of course!'

Renewable energy is undoubtedly part of the way ahead and our farm on Bodmin Moor seemed well suited to try it out. For a start, it is very windy. It lies high on a ridge in the middle of the county. From the top of the farm the Bristol Channel can be seen to the north and the English Channel to the south. There is no land mass between us and the US, as we live on the western escarpment of the moor, where steep, wooded valleys drop down hundreds of feet to the richer lowlands below. Farming on the moor has never been a very profitable business. Our farmhouse, Cabilla, is one of the five surviving Domesday manors on Bodmin Moor. A professor of hydrology started me on the road to generating renewable energy by asking to be shown the old mine workings in the valley below the house. Cornwall is renowned for its mining, especially of tin, which has been traded as far back as Phoenician times. Our woods and moorland are full of mineshafts and ancient tin-streaming beds. The Bedalder, or Warleggan, River runs down a series of waterfalls off the moor, dropping thirty metres in a kilometre just below Cabilla. A series of watercourses or leats were dug in the eighteenth and nineteenth centuries to divert water from the river to a major tin and copper mine. Powered by up

2. See   http://www.nurenergie.com/index.php/news/123/60/TuNur-files-for-authorisation-for-4-5-GW-solar-export-project

to five large waterwheels, rather than the more familiar noisy engine houses, it was called the Wheal Whisper. Abandoned for a hundred years, all the stone buildings, tunnels and wheel chambers are now overgrown with trees and creepers so that they look strangely like Maya ruins, if on a somewhat smaller scale.

The professor was excited by what he saw and said, 'But this is wonderful! You have here a ready-made hydroelectric plant. Most of the work has already been done by the Victorians. All you have to do is to clean out the top leat and you will be able to generate a lot of electricity; maybe 50 kW.'

The valley is very steep and rocky, as well as being heavily overgrown. Access is limited and I asked him if we would not have trouble installing the necessary machinery. He assured me that he had put hydro plants in much more difficult locations in places like Nepal, where everything had to be carried long distances. We began the long process of obtaining permission from all the relevant agencies. Most problematic were those concerned with migratory fish, as many of the salmon and sea trout from the Fowey River run up our much smaller river to breed. Although we received a lot of encouragement, that scheme has not yet come about, but others have and we carried on, working closely with students from the Cornish campus of Exeter University, who did some fascinating research on the various potential energy sources on our land. We installed a field array of 50kW of solar panels. Concealed by a hedge behind the house, they are completely hidden from the rest of the world. Our single wind turbine, high on the ridge, is so placed that it can be seen as a tiny object from far away, but when glimpsed from the few nearby places where it is visible I believe it is a thing of beauty. It was good to know that at last I had been able to do something practical, a change after all the years of campaigning. We now generate enough electricity to supply thirty-three average households and we are effectively off grid.

Perhaps as important as generating energy in clean ways is further improving our ability to store it. As battery technology evolves over the next twenty to thirty years, you will see more and more people moving off grid, as they will be able to capture solar power (as well as other kinds of power) with increasing efficiency and store it more effectively. It is Elon Musk who has led the way with battery technology. I installed the first two Tesla Powerwall batteries in Cornwall to store the electricity I generate. Initial costs are still a bit prohibitive, but economies of scale will change this over time, and driving our electric car is virtually free.

If international aid were to be concentrated on providing free electricity for all, just imagine how liberating the effect would be. In sub-Saharan Africa, 80 per cent of the population rely on charcoal, wood and dung for cooking. This means that the vast majority of women spend several hours each day walking to the remaining areas of forest to gather fuel. Most also lack access to clean water and 500 million currently suffer from waterborne diseases like malaria, cholera and hepatitis. If every family had solar panels, a battery, an electric cooker and a water pump, their lives would be transformed. Most already have cellphones. Ownership is as high as 90 per cent in South Africa and Nigeria, rivalling the US, which stands at 89 per cent. Africa has all but skipped the landline stage of development. Instead of hours of mindless drudgery, women especially, but also of course men and children, would be able to develop skills to supplement their incomes, to create art and artefacts and to study, maybe using the internet. Free education has never been so accessible. Using the internet, you can learn a new language or delve into the depths of metaphysics with just a click of a mouse. The web has unlocked the keys to a worldwide virtual school, potentially levelling the playing field for students around the world. And the relationship between education and a decline in fertility is widely recognised.

So my solution to reducing warfare is universal well-being through providing a level of prosperity and independence for all through switching to renewable energy worldwide, and so providing free electricity and opportunities for everyone. Alright, the electricity revolution will not in itself stop all wars, but it will show how we could all live on this earth in peace and prosperity. And I believe this can also begin to take care of the other big elephant in the global room: overpopulation.

The global population is reaching unsustainable levels. A recent report from the UN Department of Economic and Social Affairs Population Division estimates that it will reach 10 billion not long after 2050. Babies born now will be barely out of their twenties. The report says that about 83 million people are added to the global population each year.[3]

True, birth rates are at last beginning to slow down across the globe, but so are death rates, meaning more people live longer and an older population is adding massively to the increase. At present, the over-sixties represent not much more than 10 per cent of people on earth; by 2050 that will have more than doubled, so that more than one in five people will have reached retirement age, putting a huge strain on the rest of the population to look after them.

The only truly effective contraceptive is prosperity. Fertility drops as wealth rises. Paradoxically (and we seem to be the only species for which this is the case) we become less fecund as our food supply increases. The birth rate in developed countries is dropping far below replacement level. This may be asso-

3. *World Population Prospects Data Booklet* (2017 revision), https://population.un.org/wpp/Publications/Files/WPP2017_DataBooklet.pdf

ciated with female education and the growing emancipation of women. It has also been suggested that there is a correlation between the decline of fertility and the spread of television, in that people spend more time in front of the TV instead of in bed! But there is no doubt that when people know that their future is relatively secure they do not feel the need to have a large number of children to replace those who will die and to ensure the continuity of their family. One way of bringing about this sense of security would be for all people to be relieved of the daily grind by having access to free and constant electricity. It is now generally accepted that this is possible with the technology we already have, and there is no doubt that this will continue to develop exponentially over the next few years. The implications are exciting. Take, for example, the future of the motor car as we replace the internal combustion engine with cheap, silent and non-polluting electric motors. This will make the world a much nicer, cleaner, more tranquil place to live in. Even more exciting is the prospect, as self-driving electric cars take over, of not needing to own cars ourselves any more. With the ability to communicate instantly from anywhere, we will be able to summon up a driverless car wherever we are within minutes and simply tell it to take us wherever we want to go and drop us there. Private cars currently spend 95 per cent of their time parked, thus creating the need for huge car parks in every city – land which could then be released for housing and for parks for recreation: from car parks to green parks! Moreover, these clever driverless cars will, we are told, speak constantly to each other, thereby eliminating traffic jams, most accidents – and a whole lot of stress. We should be able to learn to adapt to this huge new amount of free time to develop new skills and interests, to substitute the daily strain of commuting for much more enjoyable activities. Is it too much to suggest that reducing general stress might even make us less warlike?

More realistically, free electricity could play a dramatic role in conservation. Quite apart from the huge benefit of reducing pollution of the atmosphere, it could stimulate a wide variety of methods of cleaning up the environment which are at present prohibitively expensive, largely due to the cost of electricity. Disposal of waste through pyrolysis (the thermochemical decomposition of organic material at high temperature and in the absence of oxygen) requires heating it up to as much as 500°C, but in theory everything (including car tyres and plastics) can simply be disposed of, without the costly and complicated need for recycling, and all that is left is biochar. This is a useful product which improves soil fertility, promotes plant growth and increases crop yields. Moreover, the resulting gases can be scrubbed and used to generate more electricity, and the final emissions should be much cleaner than with current methods of incineration.

Economically stable societies are less likely to go to war. The descendants of the Maya provide some good examples of how this can be brought about, even in unpromising environments.

Petén: The fall and rise of a rainforest

While Louella and I were travelling through the Maya regions of Mexico, Belize and Guatemala, we visited many temples and learned about ancient Maya history, but we also stayed with several remote communities, and it was while we were with them that the ideas I expand in this book began to take shape. Much of the time we were in the Petén rainforest, the second largest rainforest in the Americas, after the Amazon. This was the Maya's source of fuel and agricultural land. By consuming and destroying it they eventually destroyed themselves, just as we are still destroying the remaining rainforests around the world. If we had other ways of producing food and other sources of energy, we could break the cycle.

A rainforest can seem, on first acquaintance, a dull and sterile place. Unlike the great plains of Africa, where animals and birds abound, there may seem to be no life at all, just endless variations of greenery blocking the view. Nothing could be further from the truth. As W.H. Hudson recognised in *Green Mansions* (1904), probably the first romantic description of one, a rainforest is the most vibrant place on earth. Look up. Lie down and gaze into the canopy above, the cathedral roof of this amazing world. Just as the canopy of the sky was thought by the ancients to be the roof of the world, full of an infinity of planetary wonder, so the canopy of the rainforest is infinite in its variety. Far above, a golden shaft of light pierces a break in the upper foliage to reveal layer after layer of varied habitats, picking out each leaf or tuft of moss or snake-like liana and giving them a strange glory. Suspended delicately in the space between, an exquisite pattern of silver threads reveals the web of a giant tree spider. Nature makes its own architecture. In the rainforest it can sometimes take your breath away with the orderliness and balance of what is actually chaotic accident brought about by the random success or failure of individual competing species. Human artists need to worry about perspective and colour balance. Nature gets away with unplanned landscapes and jumbled layouts, which so often result in greater beauty than we can achieve, for all our artistic skill.

Although it is the richest environment on land, teeming with a diversity that defies the imagination of a European mind, the rainforest also holds creatures which are not afraid to make their presence felt. Most insects are no respecters of persons and the flying ones can be a plague, biting in clouds by day and whining singly in the ear at night. I am supremely lucky in that I do not react to bites and most insects leave me alone. Perhaps I smell bad, but I prefer to think that the truth is nearer to something I was once taught by a Maasai herdsman in Kenya. When I swatted at one of the innumerable black flies which buzzed continuously around him and his cattle, he reprimanded me. 'These small people are our ancestors and we should respect them,' he said. 'If you go gently with them, then they will be gentle with you.' And I know it works and always has for me. The less you agitate yourself and slap and scratch, the less insects will bother you.

Some birds of the deep forest, where they have not been too abused by man, will stand their ground and scream what seems like abuse as well as warning all around that there is a stranger here. There is a famous story told by Henry Walter Bates, the great naturalist of the Amazon, who one day shot a toucan to add to his vast collection of specimens. No sooner had he done so than a horde of its relations descended on him 'hopping from bough to bough, some of them swinging in the loops and cables of woody lianas, and all croaking and fluttering their wings like so many furies.' All who have walked for any time through a rainforest will have had similar experiences, though few as dramatic as Bates's. Certain birds, like the magpie-jay, act as self-appointed guardians of the forest and will often accompany one for some time, noisily announcing the impending danger man all too often represents.

Much more frightening are the howler monkeys. These are, for me, the true guardian spirits of the American rainforests. I had heard them many times in different parts of the Amazon, often far off at dawn and dusk. At Yaxja, my favourite of all the Maya sites we visited, there was a large troupe visible from time to time among the tall trees and dense forest, which has been sensitively left around the temples. Wander off the path towards one of the unexcavated mounds, where undergrowth flourishes among the faint outlines of a man-made structure, and it is soon easy to imagine that you are seeing the site as Catherwood and Stephens did way back in the nineteenth century.

The troupe at Yaxja seemed unafraid of the occasional tourist below and, foolishly thinking I might get some good pictures, I set off alone into the undergrowth towards the black shapes I could see high above. As I approached them, they seemed to take exception, leaping from tree to tree and hurling branches and leaves down on me. One gave a piercing scream and a chorus of reverberating deep voices followed immediately, rising to a terrifying crescendo. The whole world seemed to shake. No other animal on earth, I believe, can make a louder sound, although I am told that the blue whale achieves much greater decibels under water – only the human ear can't hear it! Certainly, the call of the howler monkey is far louder than the roar of a lion. Gazing up into the canopy made me dizzy and the constant waves of overpowering noise began to make me feel quite strange. For a moment I thought I was going to faint as the world above seemed to swirl around. Pulling myself together, I retreated and rather shamefacedly rejoined my wife.

The Maya Tropical Forest, or Petén, spans three countries – Belize, Guatemala and southern Mexico – and its still relatively sparse inhabitants are the descendants of the Maya. In all it is almost as big as Germany and large areas in each of the countries responsible are protected as biosphere reserves. The wet and dry seasons in the forest contrast dramatically and the resultant diversity is remarkable. More species of ant can be found in a single tree stump than in all of Europe, about two hundred species of tree and a single hectare of land may support hundreds of different species of other plants and shrubs, from vines to fungi. And, of course, as we are now discovering, all that only represents the tiny tip of an enormous iceberg of invisible life within.

The range of animals is extraordinary and this is partly accounted for by the fact that North and South America evolved separately for some 65 million years and only in the last 3 million years has there been a land bridge between the continents, allowing for the huge diversity of life to blend. As a result there are in the Maya Tropical Forest, according to a recent survey, some 163 species of mammal, 571 different birds, 121 reptiles, forty-two amphibians, sixty fish and an astonishing 3,400 species of vascular plants, that is more or less everything except mosses and liverworts.[1]

The Maya Tropical Forest is a wonderful place, one of the

great environmental treasures of the planet, and it deserves to be preserved. Fortunately, in spite of all the pressures of the modern world, much is being done and large parts do nominally have protected status. Unfortunately, as in so much of the tropical world, powerful forces constantly threaten the reserves.

PRODUCTION AND RESEARCH

'Who cuts the tree as he pleases cuts short his own life.' – Maya saying

The Maya use the same word for blood and the sap of the chicozapote (sapodilla) tree. They have always recognised the value of these trees, whose wood they used for their finest timber work, such as the lintels over doorways in the temples. We need to devise viable alternatives to cutting them down. The Maya understood their importance and made them sacred. We need to do the same, not just with these special trees, but with everything in the forest that can be harvested sustainably. That way it can produce and continue to produce vastly more than can be taken by exploiting its treasures just once for a fast buck. The added bonus is that, once saved, the Maya forest is a glorious tourist asset with almost infinite possibilities and a rare combination of culture and nature. Already Tikal is the most visited site in Central America, with up to a quarter of a million visitors a year. There are a huge number of other potential tourist attractions which could be developed throughout the area. 'Lost cities' seen in their virgin state, covered in vegetation, or just showing a glimpse of the wonders buried and yet to be excavated, are in many ways more exciting than the often excellently restored temples which abound. The special magic of visiting Maya sites, as opposed to almost all other such attractions throughout the world, is that most of the environ-

1. Cited in James D. Nations, *The Maya Tropical Forest*, 2006.

ment is still intact. This means that, if you get up early in the morning, you will hear and maybe see howler monkeys, green parrots and parakeets, a glimpse of deer and innumerable other denizens of the forest living around the temples and often quite accustomed to the presence of tourists. But remove the surrounding forest and you are left with a sterile archaeological site, bereft of its context.

The local population need viable economic alternatives if they are not to destroy the very patrimony which brings the tourists. Making and selling trinkets to them is not enough. If the tourists go away for whatever reason – and heaven knows we have seen enough reasons lately, from economic meltdown to volcanoes to airline strikes to concern about each traveller's carbon footprint – then what will they live on? Much the best answer is the harvesting of sustainable forest products.

There are so many lessons we can learn from the Maya story. We know much more now about the biology of a rainforest, about agriculture, weather patterns and how to act sustainably. And yet we are making all the same mistakes again. Deforestation, even of protected areas, is increasing. Cattle ranching is being introduced to totally unsuitable areas. Inappropriate forms of intensive agriculture are being attempted on fragile limestone soils.

So much more research is needed for us to understand the richness of the rainforest and how to exploit all its potentially priceless products sustainably. Everywhere we travelled in the Petén we saw unused research stations. Often these were barely staffed, although sometimes the facilities were excellent and crying out to be made use of.

One of the reasons scientists are reluctant to work in these remote regions is the lack of infrastructure, especially access to the internet, which makes it possible to maintain contact with researchers elsewhere. By providing free electricity, not only would it encourage vital research into still-pristine environments, but it would bolster the lives of those already living there and help them to combat incursions by those only interested in exploiting and ultimately destroying the forest.

The list of precious forest products is almost infinite and *chicle*, the original raw material for chewing gum, which is tapped from the chicozapote tree, is potentially the most valuable of them all, as the chewing gum made from it is biodegradable and preferable to the familiar sort, which is largely made of plastic. It is now being manufactured again in Mexico under the brand name Chicza and the trigger for our visit had been an invitation to come and see how it was produced. We spent some time staying deep in the forest with *chicleros*, the legendary tough men who for centuries have tapped the chicozapote trees, daily risking death from falls and snake bites as they climb up into the canopy in search of fresh sap.

CALAKMUL RESERVE

In Quintana Roo, in the Biosphere Reserve of Calakmul, we visited the reserve headquarters at Zoh Laguna. A chiclero leader, Alfonso, was with us and he sat quietly while we explained to Baltazar Gonzales Zapata, the sub-director of the reserve, what we were doing. I asked if there was any conflict between the reserve authorities and the chicleros, since I understood that in the past, especially during the first half of the twentieth century, chicle had been extracted throughout the region without restriction. At this he lit up and said how fortunate it was that we had arrived just then, as there was a presentation attended by most of his staff on that very subject going on next door in the meeting room, which we were welcome to attend. We found a PowerPoint presentation being given to about thirty people on all the problems caused by chicleros, from illegal hunting to rubbish being left in camps and inexperienced chicleros damaging the trees. After listening for a while, Alfonso exploded and took over the meeting. He had been gathering chicle for fifty-five years, he declared, and what they said was quite wrong. His people never took guns into the reserve and did not hunt there. Rubbish left at camps was a problem easily dealt with and, far from being an obstacle, the legal presence of chicleros would be a positive benefit to the reserve. They would act as guardians of the forest, as they knew and understood it better than anyone, and certainly better than some of these fresh-faced university graduates, who had never lived there. They would make sure that no poachers lasted long and that the precious chicozapote trees, most of which bore the scars of previous tapping, were looked after. The reserve should be actively harvested of its precious resources and everyone would be better off.

We found ourselves in the gratifying position of being for a moment the arbitrators between two schools of thought in what was clearly a long-running dispute. It is one which rages in many parts of the world. On the one hand, there are indigenous people who have occupied the forests for ever and gained their livelihood from them. On the other, there is a new breed of conservationists whose training tends to regard all human interference as bad. Nature should be left to its own devices and man should be removed from the equation. This has resulted in some gross abuses of human rights in the name of conservation and also some bad environmental results where, for example, traditional annual burning has maintained special habitats and, when forcibly stopped, has brought about massive forest fires. We left them discussing the possibilities of allowing some carefully monitored extraction of chicle following visits by the scientists to some chiclero camps.

I asked about the possibilities for foreign scientists to come to the Calakmul Reserve and do research. They were enthusiastic about this and said they would be very welcome to share the well-established accommodation and laboratory facilities there. They also showed us an excellent orchid garden, where many of the rare species found in the reserve were being grown. I asked if they ever had any problem with people stealing them, as I knew that they can fetch huge prices on the international market, but they assured us that this was not a problem – yet.

The lesson we can learn from the Maya example is that nature is extremely resilient and will bounce back if given half a chance. The parallel between the Maya's destruction of their environment and its subsequent recovery can give us hope that it may not be too late to start putting our damaged world back together again. It was the Petén rainforest that provided the environment from which the Maya civilisation grew – and it was its destruction by the Maya that brought about their demise.

There is a logic to having much more scientific research into such regions as tropical rainforests. Even with the superb milpa system of agriculture, there has always been a need for more. Traditionally, this has come from the forest. Meat, honey and wild fruit were there for the taking to supplement the diet. Chicle and other non-timber forest products were there to supplement the finances. Timber itself, if sustainably produced, is another source of income, but the temptation to cut down the whole forest, thereby making a fast buck, but in the process destroying the enduring capital base, is very strong and hard to resist. Only by demonstrating the real value of the forests' products will the forests themselves be saved. If ever there was a case of 'use it or lose it' this is it! The Petén is crying out for more research and I have encouraged the Royal Geographical Society and other scientific bodies to grasp the opportunity.

UAXACTUN

We visited two community forest concessions in the heart of the Guatemalan Maya Biosphere Reserve which claimed to live entirely from the extraction of forest products. Uaxactun and Carmelita were different in many ways but they both impressed us with their commitment to the forest and to harvesting it in a cooperative way. They were poor and it was a struggle to make ends meet in a harsh economic climate, but they were passionate in their belief that the forest was theirs and that they knew best how to protect and harvest it. With lots of promises from politicians and, we strongly suspected, corrupt and constantly changing government agencies, they clung to the belief that they were doing the right thing and would achieve their dreams in time. And there was no doubt that they did know a lot about what the resources of the forest were, even though they were only exploiting a few of them.

Uaxactun lies some 25 km north of the much visited Maya city of Tikal. The dirt road ends there and beyond are only rough tracks through the forest used by *xateros* (palm gatherers), drug dealers and the guards making their way to the remote and largely unused scientific research stations far to the north near the Mexican border. Normal vehicles, even 4x4s, cannot travel on these wet, rutted and overgrown tracks; only a 'bigfoot' Toyota Hilux can get through.

Wood is the most obvious forest product which should be harvested sustainably but so seldom is. Replacing its consumption as a fuel by generating renewable energy from solar power would mean that just the finest, mature timber would be harvested and so the forest kept intact. Both communities had sawmills which processed the mahogany and other valuable timber from their concessions. In the case of Uaxactun, this was an area of 83,000 hectares, of which 95 per cent is still forested.

We saw some fine planks being cut and we were assured that, with Forest Stewardship Council Certification, the trees were only taken from a designated area which would be replanted.

Both villages also had a workshop where *xate* was being prepared. This is a variety of palm that grows throughout the Maya Tropical Forest, just one of the innumerable plants covering the understorey which, to an untutored eye, all look much the same. It has been in use since Victorian times as a popular house plant. Today, the harvesting and export of the leaves to the US and Europe is one of the major cottage industries of Guatemala and southern Mexico. They are used by florists as the green backdrop of flower arrangements and are highly prized because they will stay looking fresh and green for up to three weeks.

In Uaxactun a dozen ladies were working in a large shed sorting *xate* fronds into bunches of thirty. They told us that there were two sorts of *xate*, which they called male and female. The most prized is the delicate female or *hembra* palm, *Chamaedorea elegans*, which is less than a metre high with slim, soft leaves. The larger, coarser male or *jade* palm, *Chamaedorea oblongata*, can grow to three metres. The women told us that the whole community, some 120 people, was involved and belonged to the *sociedad civil*, a sort of cooperative. 'This is our life at Uaxactun,' they said. 'We do this all year.' They get $11 for each box of thirty bunches, and the boxes are collected and transported to Guatemala City each Sunday in refrigerated trucks. From there they either continue by road to San Antonio in Texas for distribution throughout the US, or by plane direct to Europe.

While we were there, a man came in bearing a great load of cut *xate* on his shoulders. Each plant has five or six fronds on it, but they only cut one or two, so that the plant quickly recovers and, after ninety days, can be cut again. Later we were to see *xatero* camps and meet some of the men who labour in them. They all said that *xate* was becoming harder to find, as deforestation had destroyed some areas and ignorant immigrants from the highlands, who did not understand how it should be done, would take all the leaves and so kill the plant. Done properly, this is a truly renewable resource. Now they are planting seedlings, 90,000 of them so far, in a nursery near the shed.

We also saw beds of allspice seedlings there, another forest product which grows wild throughout the Petén. They are the fruit of a tree, *Pimenta dioica*, which can reach as high as 25 metres. Men climb the trees in June, July and August, using spiked boots, and then cut off the fruit-bearing branches before stripping off the berries. New branches grow and the harvest can be repeated after three to six years. Allspice, which is sometimes called pimento berry, is also grown commercially in Jamaica and has many uses. It is a key ingredient of tomato ketchup and a lot of other pickles and jams. Surprisingly, it is also used in making two liqueurs, Benedictine and green chartreuse.

Uaxactun was founded in the 1920s by chicleros to produce chicle, but since 1972 *xate* has been more important. Chicle is still harvested but, due to the lack of rain the year we were there (2009), none had been gathered by October. The villagers told us that they had six potential sources of income, mostly based on the extraction of forest products. These were *xate*, chicle, allspice, wood, looking after occasional visiting tourists and scientific researchers, and the sale of artefacts. There are problems and it is a hard life, but they were emphatic that it was what they wanted and that they would protect it with their lives. And this is sometimes necessary. The international drug industry affected them, as illegal airstrips were constructed in the forest, which made those areas unsafe to hunt or gather in. A young man had disappeared twenty days before our arrival and he was feared dead. 'Narco traffickers,' they muttered. The Mexican border lay some 50 km to the north, with nothing but virgin forest in between. The forest on both sides of the border was ostensibly protected by large Guatemalan and Mexican biosphere reserves, but they insisted that the Mexican one was full of people, some of whom crossed over to steal their *xate*. Calakmul, where we had begun our journey, lay somewhere over the border and we had seen no sign of illegal settlement there.

We stayed the night in Uaxactun in very simple accommodation, where we were able to sling our hammocks under an open round hut like a bandstand. Our host, Elfido Aldana, wanted me to understand how important their forest reserve was. The whole village was involved in it in some way; they lived almost entirely off the extraction of forest products and their survival as a community depended on it. They were doing their best to reforest the parts that had been cleared and to make the whole process sustainable. For example, where the timber operation was concerned, everything cut down was always replaced. For their efforts, they were beginning to receive some international aid. There had been grants from Germany, from Africa and from Conservation International, some of them to pay for vigilantes to help protect them against intruders, some to pay them to act as firefighters. Drug barons and *xate* thieves were the biggest problem. There are areas where it is too dangerous to go now, he said. But most were safe and they longed to encourage scientific research, both for the economic benefit it would bring and also because they wanted to understand and respect their land. He said if scientists came they would receive every possible help from the villagers.

Cocks crowed optimistically every hour throughout the night, horses tramped noisily up and down the fence beside our hammocks, trying to get at the vegetable garden, pigs squealed intermittently and a mangy pack of dogs roamed about and came to scratch for fleas beneath us.

The next day we began our journey right up to the Mexican border, which is also the border of the Calakmul Biosphere Reserve. We saw more well-established research stations at Dos Lagunas, El Cidro and Rio Azul. These all had good buildings and excellent basic facilities, but they were only occupied by a handful of guards and no scientific work was being undertaken. The problem, they told us, was that very few Guatemalan students wanted to work in the field. Once again, I felt frustrated knowing how many scientists from all over the world would give their eye teeth to be able to spend some time doing research in this rich and relatively little-studied environment.

CARMELITA

Carmelita, which we visited a few days later, was also founded as a chiclero camp, in 1925. Some chicle had been tapped the year before our visit but, as elsewhere in Guatemala, again none that year, as it had been too dry. The community's income at that time came about 30/70 from *xate* and timber. Very little allspice was being collected, as the price was too low at that time. Chicle could represent 50 per cent of the community's income, they said rather wistfully, as gathering it has a certain cachet. There was one man producing honey, another valuable commodity, as we had seen in Mexico, and there was talk of a project in the works to expand this. The difficulty, they said, was that so much was being produced in Mexico that the price was too low.

In many ways it is a model community. Carlos Crasborn, grandson of the Mexican settler of Turkish origin who originally founded Carmelita, showed us round. He was very proud of what had been achieved, but said they still faced many problems. For a start, they felt isolated from the rest of Guatemala and lacked governmental support. They were clearly dynamic and optimistic about being able to make a great future for themselves, but it was frustrating that they were so cut off from the mainstream politics of the country. This seemed to us to be a good thing, on the whole, as they avoided corruption and had an unusual sense of pride in their community. It did mean that they lacked many basic services, which communities nearer the large centres of population took for granted. Much of this could be alleviated if they had free electricity. As in Uaxactun, there was no medical centre and the nearest hospital was in Flores. Electricity came from a couple of generators and water from three small rivers and about twenty *aguadas*, ancient ponds, probably created in Mayan times. Tourism was the great hope for the future, thanks to the proximity of El Mirador, which is currently being excavated, but they still only had barely a thousand tourists a year. And, of course, there was the profitable business of servicing the scientists working at the ruins every few months. The population was stable, at about 400, and they assured me that, unlike in southern Mexico, the young people showed no great desire to escape and work in tourist hotels. We found their trust in the future and their faith that things could only get better rather touching and wished them well.

LOS GUACAMAYAS

In the west, at Los Guacamayas research station in the Laguna del Tigre National Park, we did find some good work being done, but only because of the immense energy and enthusiasm of the director, Giovanni Tut Rodrigues. He had been there for thirteen years and had virtually rebuilt the place himself. With him was a biologist called Lemuel Alfredo Valle, who spoke excellent English. The scope for research at Los Guacamayas is huge, as they were just completing new laboratories and wash-rooms and could accommodate up to forty scientists in considerable comfort. Once again, Lemuel explained, the problem was getting local scientists to come and work there. He told me that about fifty students will start the biology course at Guatemala University; about ten will finish, but most of those will go straight into the world of commerce. You can't blame them, as there is little incentive to become a field scientist, but that is a crying shame, as Guatemala has been very effective at acquiring international aid to set up these establishments.

'My country creates lots of elephants,' Lemuel said. 'The trouble is we don't have any food for the elephants.'

He and Giovanni were full of hope and plans. Los Guaca-mayas is a perfect place for research. There is one location in the park where ninety species of butterfly have been recorded, including giant blue morpho ones, and much of the area is flooded much of the time, creating extremely rich habitats. But it is the bird life which is the greatest attraction and this is also a field where there is huge potential for tourism, as they plan several bird stations where enthusiasts could come to see the 200-odd species, while basing themselves at the research station and so helping to make the whole thing economic. The iconic bird, however, and the one which has a greater draw than all the rest put together, is the one after which the station is named: the *guacamaya* or scarlet macaw, arguably the most beautiful bird in the world. There are about 300 of them living in the park, about a quarter of the remaining world population of this particular species, which is also found in a few other places: Costa Rica, Honduras, Nicaragua and Belize.

Recent research has revealed that all the macaws migrate out of the park every July/August to a place further south and come back in December/January. They go to eat the nuts of the corozo palm. These are nutritious and may have an oil which helps keep down the parasites which accumulate on their feathers during nesting. There are corozo palms in the park but they seem to prefer the ones in a big forest near Pipilas and Pontes Azules in Mexico. There they get hunted and always a few fail to return. While they are in the park they are carefully monitored and this has led to perhaps the most interesting project being worked on at the research station.

The greatest problem faced by the park, and indeed by all the protected areas in this part of the Petén, is the massive influx of outsiders who want to cut down the forest and farm it.

As always, it comes down to land rights in the end, but here there is a genuine conflict between what is best for the environment and those seeking quick profit, and new solutions need to be found.

The project Giovanni and Lemuel told us about was an exciting and original way of attempting to deal with several of these problems at once. They had involved the schoolchildren in protecting the scarlet macaws. The greatest threat to the macaws came from the robbing of chicks from their nests. These could be sold for 4,000 Guatemalan quetzals, or about $400, a fortune in that very poor area. Now seventy-five of the 125 known nests in the park are protected by a particular class in several of the schools. Each chick is named after members of the class and the children are fiercely protective of them. They use cameras to record and monitor each nest in their patch and this has led to increased numbers of chicks in some cases. They have been taught how to remove the parents gently, weigh the chicks, remove the parasites, which can present a major problem to their survival, and return them to the nest. They also clear around the trees to prevent fires and they have designed and installed some artificial nests with a lower chamber, which protects the chicks from their main predator, falcons. Bees like to occupy the nests and can also be a problem.

The results of this project have been impressive. The number of nests robbed in the last few years has dropped dramatically, the entire community now being onside and aided by the military and police, who are supportive but seldom need to take action, since the communities will deal with miscreants themselves. Even more important than the protection now being afforded to one of the world's most threatened and exquisite species has been the change in attitude by the settlers to the environment in which they now live. Involving some 3,000 children has stimulated a realisation of the significance of conservation and the economic opportunities it offers. Thanks to the schools' scheme with the scarlet macaws, a new generation is growing up which is aware of the value of the biosphere and will help to save it from some of the threats and pressures it faces.

LAS CUEVAS

The best of all the research stations we saw was in Belize at Las Cuevas. It lies at the heart of the Chiquibul Forest Reserve, the largest protected area in Belize, totalling half a million hectares, and is a key part of the Maya Tropical Forest. For ten years up to 2004 work here was coordinated by the Natural History Museum in London and a great deal of invaluable research was undertaken. Since then it has been run by a consortium of institutions comprising Maya Forest Enterprises, the Royal Botanic Garden Edinburgh and Virginia Technical University.

A trickle of good scientific work has continued and here at least there are plans for some major research. We spent a fascinating few days there with the impressive director, Nicodemus Bol, known to all as Chapal. Like so many throughout the region in his position, he was fighting a constant battle against lack of funding, recalcitrant bureaucracy and other threats. One of the most dangerous is the constant incursion of illegal *xateros* from Guatemala.

It is not so much the *xate*, which they take back across the border, as the damage they do to the wildlife by shooting for the pot, as they tend to live off the land. Chapal drove us to see a camp which had been abandoned the night before and was surrounded by empty tins, plastic bags and other rubbish. Near it we came on the 'road' the Guatemalan *xateros* had cut through the forest to bring their supplies in and the *xate* out. It was about four metres wide, a muddy track with fresh hoof marks from the horses and mules they use to carry everything. Standing up to the *xateros*, as Chapal did almost alone, was dangerous and he had had many death threats when he had single-handedly arrested them. Some soldiers of the Belizean army are stationed at the research station and a couple came with us, but they were reluctant to get involved and on this occasion stayed with the vehicle rather than risk a confrontation. They told us that some weeks before, on a training exercise, a group of cadets was fired on by *xateros*. Although equipped only with blanks, they fired back and the *xateros* ran away. When they searched the site from which they had been fired on, they found ammunition from a .22 rifle, a shotgun, an AK-47 and a GPMG machine gun!

The greatest threat to the Chiquibul Forest, though, came not from the *xateros* or deforestation, as timber extraction was fairly well managed there and had been for a long time. The real disasters, one of which had already happened and one of which was looming when we were there, came from insane plans to build massive hydroelectric schemes and so flood large areas of pristine forest. The proposal to build the Chalillo Dam on the Macal River triggered a huge campaign in the 1990s and early 2000s, which many of us became involved in. It was probably the least justified and most outrageous of all the many disastrous dams to have been built in recent years, which is saying something. The corruption and cynicism surrounding its construction were astonishing. The Natural History Museum team then operating at Las Cuevas, which is quite close to the dam site, were asked to prepare an environmental impact assessment and came to the conclusion that the impact of the dam on wildlife would be 'profoundly negative'. This report was conveniently suppressed and, in spite of all the international outcry, the dam went ahead. Robert Kennedy Jr described it as 'one of the worst cases I've seen of profiteering at the expense of people and the environment'.

Today, the Macal River, once the richest, most pure and most full of fish and other wildlife, is a polluted trickle below the dam, and the water is too dangerous to drink. Chapal took us to see the dam by a secret track through the forest. Access is strictly controlled by metal gates and barbed wire. We looked down, appalled at a winding valley which had been flooded without removing the trees, so that the water rapidly became fetid and devoid of fish and most other aquatic life. This was the last refuge, Chapal told us, of the now-rare tapir, ironically Belize's national animal. A pathetic amount of electricity is generated by the dam – the turbines were not working when we were there, due to lack of water. Moreover, the electricity could easily and much more cheaply have been provided by an alternative source of energy that would have been far less damaging to the environment, and would also have

92

boosted the country's economy. Bagasse, a by-product of the sugar cane industry, which still accounts for 60 per cent of Belize's agricultural production, could and should be used to generate electricity and would make far more sense all round. And, of course, photovoltaic (PV) and wind power are rapidly becoming the cheapest, cleanest and most efficient methods of generating electricity, especially now that battery storage is also more economical.

Not only had the colossal environmental disaster we were looking at been allowed to happen, and the worst and most gloomy of the forecasts of its effects been proved to be all too true, but a new and even more damaging dam was in the pipeline on the Vaca plateau, further up the Macal River, where the water was still pure. It was due to be constructed soon and this time the steam had gone out of the international protest, although it was an even greater disgrace.

There is a desperate need for much more research on the rapidly diminishing areas of pristine rainforest in Central America, and there are many scientists throughout the world who would gladly spend time examining the wonders of the Petén. All it takes is a bit of determination and some money. Will we really stand by and watch the same dreadful outcome as befell the Maya 1,200 years ago happen again? Have we learned nothing? There is much more at stake this time than just one civilisation. We really do seem to have reached a point where this ghastly cycle of overexploitation followed by collapse is in danger of becoming global.

Tragically, Chapal died suddenly in 2011. This was a serious blow to conservation in Belize.

It is strange to think that the fate of the whole rainforest environment may depend on what we in the industrialised world consume. If we continue to insist on the cheapest and least sustainable products to feed and clothe us and to fuel our lifestyles, then the diversity of life on earth will continue to diminish until we reach a point where our civilisation collapses. People who live in the forest understand this. Those who are lucky enough to escape for a time the hubbub of our civilisation and stand for a moment in the vibrant, busy, pullulating realm of a rainforest, or hear the utter silence of that world just before dawn breaks, can remember dimly how life once was and even gain an inkling of how it might be. But surrounded by the pounding pressures of today, whether in the clatter and roar of rich cities or in poor shanty towns, where the stench of poverty and overcrowding overwhelm those whom our civilisation has let down, we forget the lessons of the natural world. We look only to make good in the short term, and this is dangerous, for it leads to chaos and always has.

Just as with the Maya and every other civilisation that has preceded ours, this will be because we have overexploited our natural resources to the point that systems disintegrate and all sorts of unexpected, and certainly unintended, consequences kick in. Unlike those similar species to ours, rats and cockroaches, which are able to occupy every niche on earth, we will not be able to survive comfortably in the degraded world we will have created. We will take many other species down with us, but those that survive us will in a remarkably short time, a thousand years or so, restore the planet to the glorious place it was before we came along and mucked it up.

The trouble is, we will not disappear. Some humans will undoubtedly cling to life, however bad the holocaust, and in all probability they will start the whole process over again. How much better it would be for all life if we could only learn from the past and present, which demonstrate so clearly how we should behave. A good start in saving the forests of the Maya and helping their descendants is to study the products of their forests and to use them sustainably, which is also the best way to protect them.

There is no good reason why the rich resources of the Petén forest should not support self-sufficient groups and individuals who could occupy the land and harvest it sustainably. These people are the living descendants of the Maya. They struggle against corrupt government officials, predatory drug barons and the extreme hardship of living in remote parts of the Petén rainforest. A giant step towards making their lives workable would be if they had free electricity and could be independent of the cost of buying and transporting fuel long distances. Then once again the modern descendants of the Maya could flourish in the region that has so long been their home, this time showing the rest of the world how to live in harmony with their environment.

The point I make here is that these communities which at the moment both occupy and protect the rainforest are its best guardians, just as indigenous people are all over the world. If, through free electricity, their lives are made viable so that they don't get ousted by incomers and if their knowledge and lifestyles are boosted by cutting-edge research into the richness of their environment, then we can help both them and the rainforest to survive.

And, if similar local prosperity could be brought everywhere, warfare would be less likely.

Part 3

THE BLACK HORSE: FAMINE

The problem: Famine and deforestation

Here the red and black horses will be side by side. Overpopulation will not only cause War, it will also bring about Famine; and the solutions are similar. Both wars and famines can be avoided by bringing prosperity and security to everyone. One way, as I have described above, is to create a world where everyone has free electricity, and I hope I have demonstrated that this is not as far-fetched an idea as it first seems.

Famine is usually caused by mismanagement of land, such as deforestation and overgrazing. The two are linked.

If the trees are cut down, there is soon nothing left to bind the soil and it will be washed away with the next rains. As generations of farmers who have tried to grow crops in cleared rainforests have found, fertility rapidly declines since there is usually only a shallow covering of earth. Unlike in temperate climes, there is seldom deep, rich soil and crops will make erosion worse until only poor pasture land remains. Overgrazing accelerates the problem, compacting the soil, preventing plants growing and water penetrating, thus harming soil microbes and causing more erosion. A vicious circle, but one which nature, given half a chance, is extraordinarily good at reversing.

Half (yes half!) the topsoil on the planet has already been lost in the last 150 years,[1] washed away down rivers, so that waterways become clogged, fish decline and the whole system may dry up. The quality of the soil that remains changes, too, being compacted and degraded as agriculture continues. Nutrients and microorganisms decrease as they are washed out. Floods and winds move the remaining soils about and deserts are created where once was fertile land. Droughts and famine follow and people move to find new land to farm. However, this process is not inevitable and the damage can be reversed.

Much work has been done on restoring deserts and reversing their spread though tree planting and other methods. In China, on the edge of the Gobi Desert, I saw large areas of windblown sand dunes, the result of erosion, where an interesting reclamation scheme was being undertaken. Straw was dug into the sand in metre squares and then left to rot. This both binds the sand for a time and provides some humus. The idea originally was that trees would be planted later in the squares, but the system had proved to work spontaneously once the heavy labour of digging in the straw was completed. Scrub vegetation, taking advantage of the marginally improved conditions, began to sprout and a benign circle of growth slowly commenced. The infertile land next to the project was also beginning to bloom. In places, groves of trees had been planted. Where it was possible to irrigate, crops were being grown and elsewhere the land was turning green spontaneously as weeds and scrub invaded the straw-bound sand.

Schemes like this are holding back encroaching deserts in some areas and there is no doubt that reforestation helps greatly not just to bind the soil but also to achieve some small improvement in the local climate, even bringing some rain eventually.

---

1. See https://www.worldwildlife.org/threats/soil-erosion-and-degradation

But it is not enough and, due to our impatience and the desperate need to clear more land and grow more crops, even in the most unsuitable places, the situation is deteriorating much faster than it is being stemmed. It is time to consider something more radical before it is too late.

Only three things stop deforestation in the long run: national parks, biosphere reserves and indigenous land rights, where the territory is made over to its original inhabitants, who have proved again and again that they are its best guardians. To their credit, successive Brazilian governments have recognised the Indians' constitutional status as the original or first peoples, with the right to occupy and use their lands exclusively. As a result, 13 per cent of the country's land mass, nearly all of it in the Amazon basin, is reserved in 690 territories for the indigenous inhabitants. There are about 305 tribes in Brazil, apart from those still uncontacted, totalling around 900,000 people, or 0.4 per cent of the population. Putting so much land in Indian hands was a wise move, as satellite maps of the Amazon basin showing the vegetation cover today and twenty years ago reveal quite clearly that the vast majority of still-green and therefore forested areas are those which have been demarcated for conservation or indigenous control. Often the arbitrary straight lines delineating the boundaries can be seen from space. I am a trustee of a wonderful ecological reserve in the Brazilian Atlantic rainforest, the Reserva Ecológica de Guapiaçu (REGUA), another of the richest environments on earth, home to delightful creatures like the lion–tamarin monkey and where one of the most extensive bird lists anywhere has been recorded. Today, that unique ecosystem has been reduced to a long and fragmented range of forested hills sandwiched between the heavily populated coastline and the agricultural plains inland. Farmers have encroached into the hills, dissecting the remaining forests, and creating corridors across which many creatures cannot migrate. We are working hard to join up as many tracts as we can, but it is a slow and expensive business.

On Google Earth you can find maps showing the rapid rate of deforestation in the Brazilian Amazon.[2] Demarcated Indian land is shown as islands and constitutes a substantial part of what remains.

FOREST 'HIGHWAYS'

During the fifties and sixties I made the two longest journeys in modern times across South America. In 1958 Richard Mason and I drove a Jeep from Recife on the Atlantic to Lima on the Pacific, the first time a vehicle had crossed the continent at its widest point. In 1964 I went in a small boat from the mouth of the Orinoco in Venezuela to the River Plate in Argentina. Much of the then wild country through which I travelled was soon to be 'developed'. Roads were bulldozed through, settlers followed and the original Indian inhabitants massacred or dispersed. Before the road-building programme got under way, I went on another river expedition, this time in a hovercraft. With a group of scientists and journalists on board, we followed the first part of my original river journey in reverse, going from Manaus to Trinidad, which lies just off the mouth of the Orinoco estuary. One of the purposes was to demonstrate how this British invention could obviate the need for the massive destruction which smashing roads through the Amazonian rainforest would inevitably entail.

By using the existing 'highways' provided by nature, the vast network of rivers – tributaries of the mighty Amazon, along which nearly all previous penetration of the region had taken place – most roads would be unnecessary. Hovercraft, like our

2. See https://amazonaid.org/mapping-deforestation-google-earth/

thirty-seater SR.N6, and the massive SR.N4, which was then being used, before the Channel Tunnel, to transport cars and trucks from Britain to the continent, could easily provide all the transport infrastructure needed, and with very little damage to the environment. A few rapids might have to have safe passages blasted through them, but long-distance muddy tracks through pristine rainforest would not be necessary. In Finland a sleek solar-powered hovercraft has recently been developed, designed to replace much of their transport system and move over Helsinki's frigid undulated terrains covered with ice.

Once again, short-term economic forces prevailed. There was far more money to be made from contracting to build roads than there was from buying a ready-made transport system. The Amazonian forest is now less than half the size it was when I first crossed it in 1958. Recently I flew across part of that vast rainforest again for the first time for many years. The bare earth of soya plantations in the south, where often barely a tree or dried-up riverbed can be seen for miles, was as shocking as the regimented rows of oil palm plantation in Borneo, and ultimately as sterile, with little hope of subsequent recovery. Once over the still-vast forests of the north, things looked better, except that we were seldom out of sight of an arrow-straight track visible from far above, off which ominous tendrils led to new settlements. Worse still, when I looked closely I found that there was almost always smoke to be seen somewhere. Legally, or more often illegally, someone was constantly clearing the forest and burning it. The last great rainforest of the world, which had always seemed to me so vast as to be inviolable, suddenly seemed vulnerable. I was later to read that scientists at the UK Met Office Hadley Centre for Climate Prediction and Research now predict massive droughts for Amazonas, leading to collapse and devastation. They estimate that this will release immense quantities of CO_2, up to

8 per cent of all the carbon stored in the world's vegetation and soils, and a rise of 10 per cent in the ambient temperature. Global climate change would follow inevitably.

BRAZIL

Due to intense international pressure, the rate of deforestation in Amazonas has been dropping, albeit slowly, during the last decade. But now Brazil has a new president, Jair Bolsonaro, a climate change denier, whose avowed intention is, in his own words, to exterminate the Indians and cut down as much of the remaining rainforest as quickly as he can, especially that under Indian control. Two of his many quotes will suffice here:

'It's a shame that the Brazilian cavalry hasn't been as efficient as the Americans, who exterminated the Indians.'[3]

'The Indians do not speak our language, they do not have money, they do not have culture. They are native peoples. How did they manage to get 13 per cent of the national territory…? [Indigenous reserves] are an obstacle to agri-business.'[4]

Emboldened by such rhetoric, and before Bolsonaro even took office, the loggers started moving in to previously protected areas and, in 2018, illegal deforestation was up 14 per cent. June 2019 saw an alarming 88.4% increase in illegal deforestation compared to the previous year.[5] It has reached a rate equivalent to three football pitches a minute and Brazilian scientists say that it is accelerating towards an unrecoverable tipping point. All, however, may not be lost. The Brazilian Supreme Court slapped the president down when he tried to

3. *Correio Braziliense* newspaper, 12 April 1998.

4. *Campo Grande News*, 22 April 2015

5. See https://www.theguardian.com/environment/2018/nov/24/brazil-records-worst-annual-deforestation-for-a-decade and https://www.theguardian.com/world/2019/jul/03/brazil-amazon-rainforest-deforestation-environment

transfer FUNAI, the Indian affairs department, from the Ministry of Justice and place it under the control of an evangelical preacher, himself under investigation for inciting racial hatred against indigenous people; and the Indians, who are highly articulate, are fighting back, supported by massive international support.

All this just emphasises how urgent it is that more research is conducted in the remaining wildernesses of the world so that we fully understand not just what is there but what its potential value is and, above all, have the evidence to persuade governments not to destroy their patrimony.

Deforestation in the Far East has been just as bad as in South America. There were disastrous fires in Borneo in 1998 and they broke out again, even worse, in 2013. Reports came in that some 1,200 'hot spots' had been detected in Indonesia from fires, most of which were caused by the illegal clearing of forest.[6] Aircraft were grounded, shipping was disrupted and people choked from the smoke in neighbouring countries. However, that time the Indonesian government did something about it. For ninety days, throughout the worst of the dry season, cloud-seeding aircraft were sent to fly over the forests to quell the fires.

Again, in August 2018, in order to stop the smoke interfering with the Asia Games in Jakarta, 51 tons of salt were dropped over forest fires in Sumatra to stimulate rainfall, with apparent success.

The question I would like to ask is why can't these processes be used to prevent drought and famine everywhere?

6. In 2019 this figure was as high as 3,000 – see https://www.washington-post.com/business/energy/why-its-another-bad-year-for-indonesia-forest-fires/2019/09/19/9001f12a-daaf-11e9-a1a5-162b8a9c9ca2_story.html

The solution: Weather management

Everybody talks about the weather, but nobody does anything about it.
– Mark Twain

Look up and what do you see? Clouds. They are the indicators of the weather and of the state of the world. They forecast what is to come. If climate change is real, then we are going to see many strange things in the sky. The ancients used to read the stars and see their future there. Some people still do so. Isn't that strange? They seem almost irrelevant now. Stars are far distant and cannot tell us what is happening or about to happen here on earth. But clouds can. They tell the immediate future. We know so little about them, and yet they have always been feared and watched. Zeus/Jupiter, the supreme god of the ancient Greeks/Romans was the cloud gatherer, Lord of the Sky, the Rain God, the Thunderer. Apollo rode through the skies controlling the elements. The Romans had an adage, *collegit ut spargat* (collect and spread), in reference to the sun's ability to gather the clouds around it for their better dispersal, and the early Christians saw a link between clouds gathering and dispersing rain and Christianity collecting and spreading the Word of God.

Rain was the most important force of nature in the ancient world. In the Gnostic tradition, Noah did not build an ark but hid in a cloud when the rain came. Thor's hammer caused thunder and lightning in Norse mythology and its owner, who rode through the clouds, was the protector of mankind.

We have been looking too far up for signs of what is to come. We have been striving to reach out into space for solutions to our problems. Just like the ancients who studied the stars and believed that they foretold our future, we have been obsessed with the firmament of heaven and ignored what has been happening right in front of us. In every society there have been people who claimed responsibility for the weather. The belief that we could make it rain is as old as mankind itself. Only now, as we interfere with the fundamental workings of the planet in so many deleterious ways, have we developed the beginnings of a technology actually able to do so. There is a deeply felt resistance to this. We fear what dangers may be unleashed if this power falls into the wrong hands. And so we have kept it secret and, by and large, only used it so far for military purposes. It is time to grasp the nettle. We should devote proper resources to using our immense computer capacity to crack weather management for the benefit of all.

Many people, including distinguished scientists, say that the climate system is simply too complex to even consider meddling with. A familiar response to the idea is: 'We demonstrably can't predict the weather for the next week half the time and so how can we possibly forecast what it's going to do over longer periods?' But this is missing the point. Weather is chaotic locally and regionally, what scientists call an initial value problem, and is therefore very tricky to pin down. Climate, on the other hand, is a boundary value problem and much more stable globally.[1] This does mean that local weather management is more difficult than attempting to control the climate, but it should not be impossible, especially with the ever-increasing capacity and sophistication of computer technology.

Weather management has, so far, been largely the domain of the military in the various countries where it is practised. Just like nuclear power in its early days, the potential capability it could give to those who mastered it meant that it was regarded as too dangerous a subject for public consumption. In the same way that many of us who have lived much of our lives under the threat of a nuclear holocaust during the Cold War now find ourselves discussing how nuclear power may be one of the solutions to the looming energy crisis, so I believe now is the time to consider weather management for peaceful purposes.

A couple of years after the US B-29 bombers Enola Gay and Bockscar were dropping the first atomic bombs on Hiroshima and Nagasaki in August 1945, a sister B-29 from the same squadron was dropping the first silver iodide crystals over the New Mexico Desert to make rain. Many scientists, at the time and since, thought this experiment was potentially more dangerous than all that was to follow from the development of nuclear energy. There were three American scientists behind the first experiments. One was Bernie Vonnegut, the older brother of science fiction satirist Kurt Vonnegut, who is credited with discovering the potential of purified silver iodide for use in cloud seeding by creating ice and so generating precipitation. Silver iodide crystals have the same hexagonal structure as ice and can trick water molecules into latching on to them.

The second scientist was Vincent Schaefer, who invented dry ice seeding of clouds. Then there was Irving Langmuir, winner of the 1932 Nobel Prize for Chemistry, who had helped

1. See   http://www.easterbrook.ca/steve/2010/01/initial-value-vs-boundary-value-problems/ for more explanation of initial and boundary value problems in this context.

make General Electric one of the richest companies in American history by inventing an improved incandescent light bulb. He was a passionate believer in weather modification and the first to draw parallels with the power of atom bombs. He believed that it could be possible to start a chain reaction in clouds and release as much energy as had brought Japan to its knees two years before. He claimed that weather control 'can be as powerful a war weapon as the atom bomb'. The analogy was popular in military circles and that is probably why the whole subject has been surrounded with such secrecy for so long. It was perceived purely from a belligerent perspective, whereas the social and economic consequences would be far greater. The military implications were quickly recognised by the American government and a contract with the US Army called Project Cirrus was signed in 1947, which then underwrote the cloud-seeding experiments. There was great enthusiasm and a contributor to a Senate subcommittee went as far as to say, 'The nation that first acquires control of the weather shall be the leading power in the world.' The ability to produce floods or droughts over enemy territory was beginning to be seen as a new secret weapon. Soon a US government committee was forecasting that 'we could use weather as a weapon of warfare, creating storms or dissipating them as the tactical situation demands.'[2] As the Cold War and the arms race with the Russians built up during the fifties, so did a weather war. There was great optimism on the American side and it was assessed that total weather mastery would be possible in a mere four decades. That did not happen.

These experiments were but the latest in a chain of efforts to control the weather which go far back into prehistory. What else is civilisation all about if not the pursuit of happiness –

2. Quoted in Fleming (2010).

sometimes just for the elite but in real civilisations for everyone? And what constitutes happiness more than clement weather, which produces regular crops, so that all may be prosperous and content? This is the endgame, and a very old game it is. Today it is almost a taboo subject because the political implications are so scary, but it may soon be brought into play. The only way to achieve sustainability may be to start managing the weather on a global scale. There is no doubt in my mind that this is already possible.

Rainmaking: The long view

People have tried to make rain since time immemorial and it was the very essence of the relationship the Maya people had with their kings. James Frazer's monumental and hugely influential work, *The Golden Bough* (1890), started the popular vogue of looking at the customs of other cultures. He devotes a lot of time to rainmaking, since it lay at the heart of so much of almost any chief's prestige. As an introduction to an extraordinary paean on the subject entitled 'The Magical Control of the Weather', he draws examples from all over the world, from the Americas and the Far East, from Africa and from Asia.

In the section on 'The Magical Control of Rain' (it is followed by chapters on the control of the sun and the wind), he wrote: 'Of the things which the public magician sets himself to do for the good of the tribe, one of the chief is to control the weather and especially to ensure an adequate fall of rain. Water is an essential of life, and in most countries the supply of it depends upon showers. Without rain vegetation withers, animals and men languish and die. Hence in savage communities the rain-maker is a very important personage; and often a special class of magicians exists for the purpose of regulating the heavenly water-supply...

'To put an end to drought and bring down rain, women and girls of the village of Ploska [in Bulgaria] are wont to go naked by night to the boundaries of the village and there pour water on the ground...'

Today, in Romania and Bulgaria, various rain rituals, mostly pagan but some with Christian overtones, are performed in spring and during droughts. In one of them, the Paparuda, a scantily clad girl wearing vine leaves, dances through the village streets and has water thrown over her.

In 2009 it was reported that, in the Indian state of Bihar, farmers were making their unmarried daughters plough the fields naked chanting ancient hymns to embarrass the weather gods into sending rain.[1]

Frazer continues in Chapter VI: 'The foregoing evidence may satisfy us that in many lands and many races magic has claimed to control the great forces of nature for the good of man. If that has been so, the practitioners of the art must necessarily be personages of importance and influence in any society which puts faith in their extravagant pretensions, and it would be no matter for surprise if, by virtue of the reputation which they enjoy and of the awe which they inspire, some of them should attain to the highest position of authority over their credulous fellows. In point of fact magicians appear to have often developed into chiefs and kings.'

Some of the oldest records come from China and Vietnam, where kettle drums used for rainmaking ceremonies, during which they were beaten on mountain tops, have been dated to the Bronze Age. In ancient India, kingship and rainfall were divinely linked and, as in so many societies, it was the king's role to make it rain. The Australian aborigines had sacred waterholes where rainmaking ceremonies were performed. A

1. See https://in.reuters.com/article/idINIndia-41264720090723

skilled practitioner was said to be able to induce rain, at any time, to whip up thunderstorms and lightning and to divert storms away from or towards chosen areas.

These rituals and prohibitions practised by traditional societies need not be seen as superstitions, although this was how they have been regularly portrayed, especially by Christian missionaries. On the contrary, they can often be recognised as early scientific ways of dealing with natural phenomena. I remember attending a ceremony with the inland Cuna of Panama after a great draught of fish had been netted from the river. I asked to have it explained to me and was told that they were reassuring their gods that they would not be doing this again for a year. Otherwise, they pointed out, as though speaking to a child, the fish in the river would diminish. Throughout Africa and the Americas many tribes still practise rainmaking ceremonies. Many are impressive and fascinating. Some midwestern Native American tribes, like the Quapaw and Osage, became so good at forecasting the weather, probably by understanding known weather patterns, that they were able to fool traders and settlers into paying them to perform rain dances.

In 2018 President Trump nominated Sonny Perdue, a climate change denier, as his new Secretary of Agriculture. When Governor of Georgia in November 2007, while the state was suffering from one of the worst droughts in several decades, Perdue, along with lawmakers and local ministers, prayed for rain on the steps of the state Capitol. Addressing the crowd, he said: 'We've come together here simply for one reason and one reason only: to very reverently and respectfully pray up a storm... God, we need you; we need rain.' It is strange to read of a climate denier who, lacking the respect the Cuna have for their environment, is nonetheless performing what is, in essence, a rain ceremony.

PRAYERS

In Islam, *Salat ul istasqa* is a prayer performed during times of drought to ask Allah for rain. Hindus have numerous prayers to propitiate the deities and pray for rains and in traditional Buddhist cultures there are a host of worldly gods and spirits that are believed to be able to control or influence the weather. People will always be driven to prayer when in desperate straits and will continue to do so, but it is time to move on.

Scientific control of the environment accords well with modern thinking on rational behaviour and is merely a continuation of one of the most passionately and universally pursued rituals throughout history. Rather than labelling it all as superstition, it makes sense, while taking every precaution, to do what we can rather than to be afraid of 'tampering with nature'. This is what we have always done.

THE NEED FOR RAIN

Today, more than a quarter of the world's agricultural land has been degraded through overgrazing, deforestation and over-cultivation. A fifth of the global population is threatened by desertification, with all the suffering that implies. Dire droughts are becoming increasingly common around the world. Twelve million hectares of land, equivalent to about half the area of the United Kingdom, becomes useless for agriculture each year. Most recent conflicts, some four-fifths of the total, have broken out in arid areas. Currently, we are seeing the truly dreadful results of nomadic people in the Horn of Africa being driven off their traditional territory by land degradation exacerbated by climate change. They migrate to the land of other tribes who are themselves facing drought and the result is chaos, conflict and refugee camps. If soil degradation in Africa continues at its current pace, it has been estimated by the UN that soon only 25 per cent of the continent's population will be able to be fed.[2] A hundred nations are vulnerable to desertification and there will come a point when there simply is not enough soil left on earth to support us. Unless we do something about it. As Luc Gnacadja, Executive Secretary of the United Nations Convention to Combat Desertification (UNCCD), put it recently, 'The top 20cm of soil is all that stands between us and extinction.'[3] To that I would add '…and the fact that it rains.'

2. See *Africa Review Report on Drought and Desertification*, https://sustainabledevelopment.un.org/content/documents/eca_bg3.pdf
3. See https://www.theguardian.com/environment/2010/dec/16/desertification-climate-change

A short history of weather experiments

During the nineteenth century some of the first modern experiments to control the weather were tried in the US, first by burning large areas of forest 'to accelerate rain-making thermal currents'. The efforts of these so-called 'pluviculturalists' do not seem to have worked very well. More promising was the 'concussion theory' tried in Texas in the 1890s. By creating large explosions in the atmosphere it was thought that the weather would be unsettled and rain would follow. This was based on the theory first propounded by Plutarch in the first century AD that rain follows battles, and recognised by many in the military who had experienced waterlogged battlefields. The great German explorer Alexander von Humboldt had also given force to the argument by observing that erupting volcanoes caused it to rain. In Europe it had long been the practice to fire cannons at storm clouds which threatened to destroy the grape harvest with hail.

The US Congress even allocated $10,000 for experiments investigating possibilities of redirecting hurricanes and watering the arid lands of Arizona, but they proved unsuccessful. However it did spark a lot of satire in contemporary literature, notably Mark Twain's book *The American Claimant*, published in 1892. In this he has a character, Colonel Mulberry Sellers, who plans to control the world's climates and sell them for cash, 'taking the old climates in part payment, of course, at a fair discount, where they are in a position to be repaired at small cost and let out for hire to poor and remote communities not able to afford a good climate…' He also intends to supply 'fancy brands for coronations, battles and other great and particular occasions'. He would probably have been amazed by the British travel company that offers a 'cloud bursting service that can 100% guarantee fair weather for your wedding day', with photographs, for a mere £100,000.[1]

There has been an ideological angle to research into weather control too, just as there was in the space race, and there was competition between the great powers to come up with the most impressive schemes. Before the end of the Cold War, the Russian government had developed techniques, for both military and peaceful purposes, of making it rain by seeding clouds for major celebrations. They claim that they order up clear skies for Moscow's May Day parade each year and it seems it never does rain then. During the 2005 celebrations in Russia, which were attended by President Bush, silver iodide, liquid nitrogen, dry ice and even ordinary cement were dropped from between 3,000 and 8,000 metres about 100 km from Moscow by twelve aircraft and the result was that the rain, which had been falling up to five minutes before the parade, cleared and allowed an impressive fly-past by Sukhoi and MiG fighters. This was done on President Putin's express orders.

Thailand has a bureau in charge of cloud seeding with 600 employees and a budget of $25 million. It was inaugurated in

1. https://www.oliverstravels.com/blog/guarantee-perfect-wedding-weather-olivers-travels/

November 1955 by King Bhumibol and is run by the *Department* of Royal Rainmaking and Agricultural Aviation.[2]

Most cloud seeding in Thailand has been performed to bring rain to drought-affected farmers, but in 2018 the technique was used, apparently with some success, to seed clouds over Bangkok in order to reduce the very high atmospheric pollution in the city at that time. In 2009, Jordan received permission from Thailand to use the technique. This is perhaps the only example left in the world where a reigning monarch has followed the ancient tradition of being his people's rainmaker!

In 2009 it was reported that Indian scientists were developing their own technology to seed monsoon clouds. They have been flying in a light aircraft into storm clouds measuring the temperature, speed, chemical composition and moisture and particle levels of the clouds from the inside. This was part of a new plan to seed the clouds with rain-inducing chemicals to try to control the timing of the annual monsoon, whose late arrival was causing havoc that year. Two billion people in Asia depend on monsoon rains to grow their food. If the rains fail, they will starve.

The Chinese, who a Greenpeace report says will have an inadequate food supply by 2030,[3] are said to spend $40 million a year on weather management to alleviate droughts and stop hail, which would otherwise damage crops. Around 35,000 people are employed in this work and they used it enthusiastically to create good weather for the 2008 Olympics, as was widely reported in the press at the time.[4]

---

2. See http://www.tsdf.or.th/en/royally-initiated-projects/10167-royal-rainmaking/

3. http://www.greenpeace.org/eastasia/press/releases/climate-energy/2008/food-security-climate-change/

4. Reported in *BusinessWeek*, 24 October 2005, for example.

MILITARY WEATHER MODIFICATION

There is a history of countries using weather modification for military purposes. During the Vietnam War, between 1967 and 1972, the US spent $12 million on seeding clouds above the Ho Chi Minh trail in order to induce heavy rain to create landslides, wash away river crossings and enemy crops, reduce traffic through the jungle and so disrupt Viet Cong supply lines. There were two operations, still top secret, codenamed Project Popeye and Project Motorpool, which flew thousands of missions out of an airbase in Thailand without the knowledge of the Thai government. No scientific data were collected and so it is hard to say how successful they were, although as a PR exercise they proved disastrous. When, during the Nixon era, information started leaking out, it became known as the Watergate of weather warfare. The defence that it was better to 'make mud not war' fell on deaf ears. Better work appears to have been done in seeding the whole of the Philippines for drought relief in 1969.

The French had tried the same thing when their forces were besieged at Dien Bien Phu in 1954. They attempted to do this by shipping 150 baskets of activated charcoal and 150 bags of ballast from Paris in order to make artificial rain. In 1963 the CIA are reported to have seeded clouds over demonstrating Buddhist monks. They had found that the monks resisted tear gas but ran away when it rained.

There has for some time been concern about these sorts of weather management being used for military purposes. In 1977 the United Nations Convention on the Prohibition of Military or Any Other Hostile Use of Environmental Modification Techniques (ENMOD) was adopted by the UN General Assembly. Since then, eighty-five nations have ratified it. In the Introduction to the Convention, it recognises the possible benefits and lists: 'Realizing that the use of environmental modification techniques for peaceful purposes could improve the interrelationship of man and nature and contribute to the preservation and improvement of the environment for the benefit of present and future generations.' It also goes on: 'Recognizing, however, that military or any other hostile use of such techniques could have effects extremely harmful to human welfare' and then later it specifies the sorts of phenomena it is talking about.[5] These include earthquakes; tsunamis; changes in weather patterns (clouds, precipitation, cyclones of various types and tornadic storms); changes in climate patterns; changes in ocean currents; changes in the state of the ozone layer; and changes in the state of the ionosphere.

Military use of these techniques is prohibited, but there is evidence from recently declassified British government papers that, at least during the Cold War, the US and Russia had competing secret military programmes aimed at controlling the world's climate. The *Daily Express* reported in 2007 that one scientist had boasted, 'By the year 2025 the United States will own the weather'. Certainly weather manipulation could be a terrible weapon in the wrong hands, but if the research is already under way, surely it makes sense to find out what is going on and, perhaps, to use the information already garnered as the basis of research which might really benefit mankind.

CLOUD-SEEDING EXPERIMENTS

In the US, $20 million a year was until recently being ploughed into cloud-seeding research, still peanuts compared to NASA's $20 billion budget for space research. In many states a variety of operations are carried out annually. Notably, ski resorts pay

---

5. https://fas.org/nuke/control/enmod/text/environ2.htm

companies to tease more snow from the clouds and Vail Ski Resort in Colorado estimates that doing this increases its snow-pack by 15 per cent. The techniques used have all been based on 'milking' clouds by sprinkling them with silver iodide. The particles are supposed to cause the clouds' ice crystals to accumulate in clumps, which are too heavy to stay aloft and so fall as rain or snow. However, scientists admit that they cannot fully explain how the seeding process inside clouds works. Quite a lot of research is being undertaken, but not nearly enough.

Things can also go wrong. At the end of October 2009, a year after they had hosted the Olympics, the Chinese were embarrassed when their scientists triggered a heavy snowfall over Beijing. Ahead of an advancing cold front, 186 doses of silver iodide were fired by rocket into the base of the approaching storm clouds during a Saturday night, causing an extra 16 million cubic metres of snow to fall on the city. Zhang Qiang, the deputy director of the weather modification office, told state media: 'We won't miss any opportunity of artificial precipitation since Beijing is suffering from the lingering drought.[6] She justified their action by asserting that, when melted, the snow would bring great benefit to local farmers, as an area of some 800,000 hectares of farmland in the area was suffering an unusually severe drought. However, international travellers were less pleased, as around a hundred planes were delayed and some cancelled at Beijing airport.

We have seen that the very first experiments in dropping dry ice were undertaken by Irving Langmuir as part of Project Cirrus in 1947. He was attempting to divert a hurricane, which had caused a lot of damage in Miami before swinging 350 miles out to sea and drifting gently northwards. A US Air Force B-17 was used to sprinkle frozen carbon dioxide (dry ice) from 500

6. Sic, https://www.theguardian.com/world/2009/nov/02/china-snow-beijing

feet above the eye of the storm before circling for a while and then turning for home. For reasons which seem never to have been explained, the hurricane followed, gathered momentum and slammed into the coast near Savannah, Georgia. Several people were killed and many millions of dollars worth of damage were caused. Although the military denied any involvement, lawsuits were threatened and conspiracy theories, for once perhaps realistically based on something sinister, raged for many years. Although Langmuir was soon after to state that 'There is a reasonable probability that in one or two years man will be able to abolish most damage effects from hurricanes,' this failure caused everyone to be nervous about tangling with hurricanes, a caution which continues today after over seventy years. The damage they and tornados do annually in the US alone is vast and yet research into controlling them is minimal. Project Cirrus was abandoned and the cause of seeding hurricanes was set back by over a decade.

Project Stormfury was another US government project, which ran from 1962 to 1983. It attempted to weaken tropical cyclones and hurricanes by seeding them with silver iodide. There was great enthusiasm at the time from those involved. One scientist likened their activities to big game hunting, perceiving hurricanes as 'the largest and wildest game in the atmospheric preserve. Moreover, there are urgent reasons for 'hunting' and taming them.'[7] Even today, the emphasis in weather control is more on controlling hurricanes than anything else. Bill Gates is on record as saying that he intends fighting hurricanes by manipulating the sea, 'draining warm water from the surface to the depths, through a long tube.'[8] Gates is also concerned about the impact climate change will

7. Quoted in Fleming (2010).
8. *Business Insider*, 16 July 2009.

have on poor people in the developing world and he has donated funds for research into geoengineering, which we will come to in Part 4.

Recently, in August 2019, President Trump was reported as saying in a meeting with top national security officials about the threat of hurricanes: 'I got it. I got it. Why don't we nuke them? They start forming off the coast of Africa, as they're moving across the Atlantic, we drop a bomb inside the eye of the hurricane and it disrupts it. Why can't we do that?'

Two days later, he denied having said it and claimed it was 'fake news'; but the issue is again current.

In the UK in 1952 experiments called Operation Cumulus were conducted from Cranfield in Bedfordshire, in which clouds were seeded from aircraft. On 15 August, in good weather, one pilot later reported spraying some clouds with salt and being told with his colleagues, some of whom may have been using silver iodide, that the rain had been the heaviest for several years and all out of a clear blue sky. They cheered and toasted meteorology until later that evening, when they heard the BBC announcement of the Lynmouth disaster, and 'a stony silence fell on the company'. Nine inches of rain fell on the East and West Lyn rivers in twenty-four hours, bringing 90 million tons of water down the narrow valley, destroying buildings and taking thirty-five lives. News reports described it as the most destructive storm in British history. The Met Office and Ministry of Defence denied that there had been any experiments, but when the files were examined thirty years later all the relevant ones were missing. Operation Cumulus was closed down soon afterwards.

After the Chernobyl disaster in 1986, heavily radioactive clouds began to build up and move north with the prevailing wind towards Moscow and St Petersburg, then still called Leningrad. A massive cloud-seeding operation caused heavy, contaminated rain to fall in rural Belarus, thus avoiding catastrophe for millions. However, the Soviet authorities chose not to warn the population where the rain was made to fall and as a result the area was devastated and there have been many cases of leukaemia, cancers and birth defects since. This could easily have been avoided if people had been issued with potassium iodide tablets and told to stay indoors.

Despite the risks inherent in manipulating the weather, geo-engineering offers the best chance of meeting the climate challenges that lie ahead. The systems governing our planet are extremely complicated and it is essential, if we are to survive, that we devote every resource to trying to understand them. What we should not do is allow a superstitious dread of nature to stop us doing what our brains tell us is possible. As long ago as 1945, a thorough discussion of the subject was published in the US by Vladimir K. Zworykin, research director at the Radio Corporation of America (RCA) Laboratory. He concluded that there should be much more research into meteorology, leading to an international organisation with a rapid-deployment force ready to intervene in the weather as it happened, by whatever means was available. He wrote: 'Such an international organization may contribute to world peace by integrating the world interest in a common problem and turning scientific energy to peaceful pursuits. It is conceivable the eventual far-reaching beneficial effects on the world economy may contribute to the cause of peace.'[9]

John von Neumann, who worked with Robert Oppenheimer to create the atom bomb, is known as one of the brightest minds of the twentieth century and as the godfather of modern computer programming. He believed that, after splitting the atom, there remained three great scientific break-

9. Quoted in Fleming (2010).

throughs to be achieved: understanding the brain; understanding cells (genes and DNA); and, his pet project, conquering the physical environment, which he called 'to jiggle the planet'. He invented numerical weather forecasting and claimed that one day computers would be powerful enough to predict and control the climate.

SEEING THROUGH THE FOG

It was Captain Robert FitzRoy of Beagle fame who wrote the very first public weather forecast. It was published in *The Times* on 2 August 1861 and began: 'General weather probable in the next two days.' To begin with, forecasting was mainly for shipping and, of course, a significant part of it still is. Only with von Neumann in the late 1950s did computers start to be used, which were crude at first compared with today's supercomputers. These, I learned on a recent visit to the amazing British Met Office HQ outside Exeter, can now do 710 trillion calculations per second – that's twelve zeros. Von Neumann's day has now arrived.

Climate and weather are different, of course. Climate is the long-term average pattern of weather for a particular region, the aggregate of weather conditions over time. Weather is a snapshot of atmospheric conditions at a specific place and time.

Climate management is already well accepted. We manage the climate in a negative way every time we cut down some more rainforest or release CO_2 from the chimney stacks of coal-fired power stations. We manage it positively whenever we plant new forests or attempt to reduce our CO_2 emissions by, for instance, implementing the Montreal Protocol and so continue to phase out the production of chlorofluorocarbons, which are responsible for ozone depletion. This was the only legally binding environmental agreement signed by all countries in the world and the most successful environmental effort in human history.

Although weather management is expensive, it has already worked sometimes in recent decades. During the Second World War, after the Battle of Britain, it was recognised that fog was a major obstacle to aviation, delaying bombing raids and causing many accidents. The Germans were said to be experimenting with fog-making machines to confuse Allied aircraft. The British developed a method of clearing fog called Fog Investigation and Dispersal Operation (FIDO), which has been described as one of the most spectacular but least publicised secret weapons of the war. Runways had open trenches dug alongside them and were either filled with fuel or lined with pipes delivering petrol, which when ignited raised the ambient temperature and dispersed the fog. Huge quantities of gasoline were consumed, as much as 100,000 gallons an hour, but FIDO has been credited with saving the lives of up to 10,000 airmen and it has been suggested that it helped to shorten the war. The brilliant engineer behind FIDO, Guy Stewart Callendar, was also the first scientist to recognise the contribution of fossil fuels to the greenhouse effect, but his invention was abandoned after the war, as radar and other instrument landing methods replaced it.

A much more exciting use for fog has also been developed in recent years. It is called fog harvesting. In many of the driest parts of the world, like the Atacamo Desert, there is little or no rain, but fog rolls in daily from the ocean to dissipate far inland when it reaches a mountain range. By erecting large polypropylene plastic nets like vast beach volleyball courts, as much as two-thirds of the tiny water droplets in the mist are trapped as they drift inland. They then trickle down to a reservoir to create a regular supply of fresh water. A 48-square-metre net, costing about $400, might produce as much as 750 litres of pure drinking water a day,[10] although with current technology 200 litres seems to be nearer the mark. This may not be enough for irrigation, but it is highly significant to a community, who will otherwise have to walk many miles to fetch often contaminated water and carry it home. Experiments have been carried out in many parts of the world where conditions are right. They have proved exceptionally successful in Oman, where on foggy days the yield was up to ten times that achieved in the Atacamo. There must be places all along the arid coasts of Africa where such technology could be introduced at minimal cost but with real long-term benefits. So much better than hugely expensive and ultimately futile refugee camps.

This technology is not new. Pliny the Elder describes a 'Holy Fountain Tree' growing on the Canary Islands where until quite recently people harvested fog droplets from the leaves of trees on the island of El Hierro. Nature, too, is adept at extracting water from the atmosphere and could teach us how to do it better ourselves. Many creatures do this in a variety of ways, often using dew. One, which has led to pioneering research, is the darkling beetle which lives in another of the harshest environments on earth, the Namib Desert. Microscopic bumps and troughs on its back condense minute water droplets from early-morning fog and channel it down to the beetle's mouthparts. Now, using plastic sheeting with the same hexagonal design as that on the beetle's back, research companies are designing improved methods for condensing liquid from a vapour. One is the Dew Bank Bottle, which won the Bronze Prize at the Idea Design Awards 2010.[11] Left out

10. See the open source water and sanitation resource Akvopedia, https://akvopedia.org/wiki/Water_Portal_/_Rainwater_Harvesting_/_Fog_and_dew_collection_/_Fog_collection_and_storage
11. See http://www.yankodesign.com/2010/07/05/beetle-juice-inspired/

overnight, it will provide a glassful of fresh water to drink in the morning. In another experiment, tiny glass spheres were embedded in warm wax to create a pattern of water-attracting peaks and water-repelling troughs, a system which it is hoped will be several times more efficient than current water-collecting methods.[12] Apart from the obvious value to desert campers, the potential for providing water in congenitally drought-ridden regions is immense. How much better to collect the purest possible water in this way, rather than to go to all the labour involved in digging wells or watercourses, which are all too prone to contamination.

Desalination is a well-known, if rather inefficient, method of obtaining usable water as it requires a lot of energy. Reverse osmosis, whereby sea water is forced repeatedly through a membrane to remove the salt, is now overtaking distillation, but both methods are expensive. A new technique, which is being developed by Germany after research in the Negev Desert in Israel, shows promise. Although a desperately dry place, the annual average relative air humidity in the Negev is 64 per cent, which means that in every cubic metre of air there are 11.5 millilitres of water. Using PV panels to pump saline water running down the outside of tower-shaped units, water is attracted from the atmosphere, boiled and run off into storage tanks.[13] Drinking water is thus obtained from air humidity at zero cost.

As there is far more water in the atmosphere than in all the rivers of the world, some other new techniques for extracting this are also being tried. One, in Mexico, is using Russian technology. This is ionisation. Charged ions are generated on the ground and pointed at the sky. The theory is that this

12. See http://news.bbc.co.uk/1/hi/sci/tech/1628477.stm
13. See https://www.sciencedaily.com/releases/2009/06/090605091856.htm

causes dust particles and ice crystals touched by a charged ion to clump and fall in the same way as rain. There is still much scepticism about this, but it does seem to work and the rainfall in northern Mexico and Baja, California, where there are now thirteen ground stations, is claimed to have increased by 30 to 35 per cent. There are signs that regular cloud seeding may soon become respectable. With new radar technology, advanced computer modelling and much faster supercomputers, scientists should be able to begin to unscramble the fiendishly complicated way in which weather patterns form, and start to control them. The American National Academy of Sciences did call for a concerted national research programme in 2003, but so far there appears to be no federal funding for it.

Another Swiss team has been using lasers for the first time to cause condensation outdoors. As of last year they could only create condensation along the laser channel, but they are now working on condensing a wider area by sweeping the laser across the sky.[14] This has been described as potentially 'breakthrough technology'.

In Abu Dhabi a secret $11 million weather-modification project arranged by a Swiss company, Meteo Systems International, and monitored by the Max Planck Institute, has built fifty ion-emitter towers 10 metres high topped by a two-metre-square electric grid, which generates negatively charged particles to send dust into the atmosphere. The hot desert air helps send the negatively charged particles high into the sky by convection. The theory is that, at the right altitude, the particles attract dust, around which water molecules in the atmosphere condense. The longer these stay at this altitude, the

14. Joly P., Petrarca M. et al., *Appl. Phys. Lett.* 102, 091112 (2013); https://doi.org/10.1063/1.4794416

more chance of clouds forming, and then water droplets. The researchers claim to have created fifty-two unanticipated rain showers out of seventy-four experiments, always out of a clear blue sky on days when no rain was forecast.[15] The process, if it works, would be cheaper than desalination and might solve the chronic severe drought in the region...

The four main areas in which benign weather modification (BWM) has been tried are: making it rain to alleviate drought; suppressing hail to protect crops; dispersing fog to improve airport efficiency; and making it snow to improve skiing. Of these, it would seem to me that creating rain is the one with far the greatest potential benefit.

Water is rapidly becoming a major world issue and the cause of conflict. Water resources everywhere are being strained as underground aquifers are drained and populations grow. International conflict over water rights looms as countries vie for dwindling resources. There are 214 river basins around the world shared by more than one country, and most rivers, especially in Africa and Europe, are multinational. This makes it all the more urgent that ways are found to squeeze more rain out of clouds through BWM.

15. Reported in https://www.arabianbusiness.com/abu-dhabi-backed-scientists-create-fake-rainstorms-in-11m-project-371038.html

The politics of population

As I write this, news is coming in of what is being described as the worst drought in the Horn of Africa for sixty years, perhaps ever, developing. Once again urgent appeals are being made for us to help alleviate the terrible suffering it is causing. I have the greatest admiration for welfare workers and on many of my travels I have worked alongside them and been impressed by their commitment and dedication. It is an admirable thing to devote one's life to helping others, often for far less reward than one's talents would earn elsewhere. However, troubling thoughts lurk in the back of my mind when I consider the issues and possible solutions. They are there because I am and always have been far more interested in real, long-term solutions to the world's problems rather than quick fixes. Of course all our heartstrings are tugged when we see images of starving children with wide eyes, and their mothers who sacrifice everything for their survival. And of course we dig as deep as we can in our pockets to give them instant help. But is this really the best way to give them and the millions like them a future?

By delivering food to drought-stricken areas are we undermining the ability of the farmers to recover and rebuild their agricultural economy? By establishing facilities for the aid workers, where they receive adequate food and clean water and have hygienic living conditions, as must be done or they would themselves not survive or not go there, are we not creating a milieu of 'them and us', while presenting examples of unattainable standards of living?

It would be so much better if we could prevent drought in the first place. Much the fastest and most effective way to do that would be to make it rain. If we have the ability to do this, then it is criminal not to make every effort to bring it about. After all, human life on earth is wholly dependent on water, both for drinking and irrigation. Without it, we are doomed anyway, whatever happens to the climate.

The evidence that the earth is warming is now unequivocal. The impact of that warming is already being felt and we need to do something about it. It has been a human trait throughout our history that, if something is possible, sooner or later we are going to try it, come what may. It must make sense to investigate this particular Pandora's box rationally and while the technology is still young rather than leave the debate until global warming becomes a major source of global conflict.

Economics will always determine behaviour. Therefore, the only way to solve the world's impending food crisis is to make everyone more prosperous. The easiest way to do that is to make sure everyone has enough to eat, everywhere. And the best way to do that is to start managing the weather. I believe there is still time, but time is running out. Too many people have fiddled while our planet burns and complacency looks set to kill us all in the end. It may already be too late, but we must try. We are the only species capable of putting things right. It is time to stop arguing about who and what is to blame and to face up to our responsibilities. And we are quite good at doing this when it comes to the crunch, although the way is often rocky and nothing comes easy.

POPULATION GROWTH

At the start of the Classic Maya period, there were barely 200 million people on earth. A thousand years later, in 1804, global population reached the billion mark. In 1927 there were 2 billion, 3 billion in 1960, 4 billion in 1974 and 5 billion in 1987. By the year 2000 global population had risen to just over 6 billion and the Intergovernmental Panel on Climate Change (IPCC) made three world population projections for the year 2050: 8.7, 9.3 and 11.3 billion. Already, nineteen years later, global population has passed the 7 billion mark and so the larger prediction now seems the most likely. There is increasing concern that the world has undergone a permanent paradigm shift, whereby the number of people on earth has outstripped the planet's ability to support them. James Lovelock believes, and many agree with him, that already with a global population of 7 billion, life as we know it today is not sustainable. As ever more humans consume ever more resources, the situation can only get worse. Something has to give if we are to avoid mass starvation and find a way of producing much more food.

Population growth is certainly one of the major problems facing us and it is one where huge strides could be made if there were but the will. The Chinese government has calculated that its 'one-child' policy has brought about, by the avoidance of some 300 million births, a reduction of about 1.2 billion tons of CO_2 being emitted annually into the global atmosphere. This is a greater reduction than has been achieved by all the measures of the Kyoto Protocol. However, population remains the Cinderella of the great sustainability debate and CO_2 emissions continue to rise by over 5 per cent per annum. The reason is that tackling population growth is fraught with political, religious and social difficulties. Conservative attitudes, especially when based on religious prejudice, tend to oppose contraception and abortion and favour growth. Liberal opinion usually supports human rights and abhors coercion, and is thereby reluctant to endorse compulsory population policies. As a result, there is stalemate, which just allows population growth to continue.

Thomas Malthus, in his *Essay on the Principle of Population*, said in 1798 that 'the power of population is infinitely greater than the power in the earth to produce subsistence for man. Population, when unchecked, increases in a geometrical ratio.' He went on to state that agricultural production increases arithmetically, i.e. much more slowly, and therein lies the biological trap that humanity can never escape. This fact lies at the heart of the human problem and we seem incapable of resolving it. Doing so is about much more than exercising self-control personally, nationally or even globally. There is an irresistible urge to reproduce built into our most fundamental DNA, as it is in every species that has ever lived. Until we find a way of overcoming this suicidal imperative, let us see if we can find a way of helping our fragile planet, with its limited resources, to support us. Selfish, yes; but as Richard Dawkins pointed out some time ago, all genes are selfish. Darwin wrote: 'There is no exception to the rule that every organic being naturally increases at so high a rate that, if not destroyed, the earth would soon be covered by the progeny of a single pair.'[1] To which Dawkins responded: 'It is differences that matter in the competitive struggle to survive.'[2] The big difference about us

1. *On the Origin of Species*, 1859, III.
2. *The Selfish Gene*, 1989, p.37.

is that we are perfectly capable technically of solving everything. We just appear not to be sufficiently evolved to agree on the correct course. For example, there is a glaring dichotomy between the concern we all have at the apparently unstoppable overpopulation of the world and the extreme distress, often whipped up by the media, over each single life lost publicly and each one saved. Every life is deemed precious and yet we all know there are too many of us.

The solution is for women to be empowered to decide how many children to have; but they will tend only to choose to have fewer children if they are secure and relatively affluent, not at daily risk of losing their offspring to famine or disease, often caused by unreliable weather. Prosperity leads to falling birth rates. People's very prosperity and security will then do more than anything else to reduce carbon emissions and so contribute to combating climate change. It has been calculated that an American woman deciding not to have a child will decrease her carbon legacy by about 9,441 metric tons of CO_2, thereby lowering her own cumulative carbon footprint by more than 80 per cent without any other lifestyle changes.[3]

Change is coming all over the world and the most visible alteration will be in the weather – not only the sort of modification which is the result of the immediate local influence people have had on the climate through cutting down forests and creating deserts. This is the result of something much greater and more apocalyptic. It is global heating and we ignore it at our peril.

From now on, humanity can only go in one of two directions. Either we hark back to a golden age and try to recreate it from the ravages we have wrought; or we use our ingenuity to

3. See   https://www.biologicaldiversity.org/programs/population_and_sustain-ability/pdfs/OSUCarbonStudy.pdf

make something that is rapidly becoming unsustainable work again. Going back to living closer to nature is a lovely idea, but it is simply not practicable for most people today. The comforts of modernity – unlimited energy, fresh water, rapid transport and clean clothes – have become too familiar and seductive for many to abandon them willingly. The bottom line is becoming increasingly apparent: either we head on blithely towards our own extinction or we take control of the climate.

Conflicts over water, often described as the new oil, are becoming increasingly prevalent and we have not seen the last of them. Many countries are so dry that their economies depend on other things than safe, dependable agriculture. From Afghan poppy crops to Yemeni terrorism, constant drought causes problems. Worse when, as in the Middle East, life–giving rivers such as the Jordan, Tigris and Euphrates cross international boundaries. Drought causes famine, which in turn causes mass immigration. Or, as in the case of Somalia, those who remain are forced to turn to piracy, although in that case the arrival of large fishing vessels from the EU and elsewhere with massive nets, which take all their fish, is another cause. Throughout virtually the whole world, water is becoming an increasingly contentious subject and the demand for it is accelerating everywhere. Sometimes this is the result of prosperity, when people demand more water for baths and showers, lawns and golf courses. More often it is the result of poverty, as rivers dry up and people have to walk ever further to collect water; or as wells and boreholes are sunk ever deeper, the water table drops and people and animals die of thirst.

Manipulating the weather is achievable today and the budget required to do it would be peanuts compared with the astronomical costs of famine relief, let alone of the wars in which water so often plays at least some part. By guaranteeing reforestation programmes through ensuring rainfall at significant times in the growing season and by making deserts – which today constitute a third of the world's surface – bloom, we would be able to do more to combat global heating than all our feeble efforts to insulate our homes and generate electricity can ever achieve. We are already suffering from the effects exploding population, uncontrolled pollution, deforestation and industrial practices are having on the climate. It is time for us to take matters into our own hands and start to restore equilibrium. There is plenty of water out there – it just tends to fall in the wrong places. If the technology of making and stopping rain develops fast enough, it need not be a competitive business and only be available to the rich nations. There is enough water stored in the clouds to irrigate everywhere on earth amply, only at the moment most of it falls into the sea. Would changing this matter? Would there be side effects from the diminution of fresh water landing on the salt? Perhaps, but that might be a price worth paying for vastly improved agricultural yields everywhere. It is probably also instantly reversible if things went wrong. Slowing the process seems unlikely to leave any after effects but, of course, we don't know this and as with all radical initiatives, there are risks.

We live off the land. We always have. It provides our food, our shelter, our warmth and our clothing. It makes every kind of sense to cherish the land and nurture it for future generations. This is what good farmers and land managers do. Today, management has morphed into exploitation, but then we have always had to balance our needs against what the earth could produce. We turned the earth to make crops grow. We dammed and diverted rivers for irrigation. We hunted the forests and plains for meat and we trawled the lakes and oceans for fish. Although it is clear to everyone that we are doing all these things to excess, there is little or no indication that we are prepared to slow down, in spite of huge efforts by scientists and environmental campaigners to encourage us to do so. And the population grows and grows. If we are unable to consume less, then we have to find a way of producing more. The Green Revolution, which is claimed to have saved a billion people from starvation, appears to have run its course. For the last few decades new strains of rice and wheat have delivered rapidly increasing crop yields, which have just about kept pace with population growth in some areas, but they have been matched by accelerating desertification. The time has come for another solution if we are to avert catastrophe. The logical answer is weather control. For all its dangers, it is the only way left to us. If we can do it, and it seems we can, then one day we will. That being the case, let's get on with it, because the sooner we master the technologies required to bring it about, the sooner we can start alleviating the damage we have already done.

POLITICAL WILL

What possible chance is there of our divided world agreeing where the rain should fall? Yet this is exactly what we have been urged to consider ever since the UN Framework Convention on Climate Change was signed in 1992. In Article 2 it states: 'The ultimate objective of this Convention… is to achieve… stabilization of greenhouse gas concentrations in the atmosphere at a level that would prevent dangerous anthropogenic interference with the climate system.'[4] Many believe that we have already passed the point where our dangerous interference can be halted, let alone reversed, and that now the only course of action left to us is to attempt to put right the damage we have caused. All else are counsels of despair.

John F. Kennedy, addressing the UN in September 1961, shortly after Yuri Gagarin had become the first man in space,

4. http://legal.un.org/avl/ha/ccc/ccc.html

recognised the significance of weather control being in the same league as the space race and satellite communications. It is worth quoting what he said, as this was part of one of the most important speeches he ever made, just two years before he was assassinated:

'And as we extend the rule of law on earth, so must we also extend it to man's new domain, outer space... To this end, we shall urge proposals extending the United Nations Charter to the limits of man's exploration in the universe, reserving outer space for peaceful use, prohibiting weapons of mass destruction in space or on celestial bodies, and opening the mysteries and benefits of space to every nation. We shall propose, further, *cooperative efforts between all the nations in weather prediction and eventually in weather control* [my italics]. We shall propose, finally, a global system of communications satellites linking the whole world in telegraph and telephone and radio and television. The day need not be far away when such a system will televise the proceedings of this body to every corner of the world for the benefit of peace.'

Prescient stuff! In my opinion much of the vast amount of money and effort put into space research has been wasted. The figures are quite literally astronomical. On Thursday 22 July 2011 the Shuttle Atlantis touched down at Kennedy Space Center for the last time. Its thirty-three flights and the hundred other shuttle flights originally planned as part of a programme to reach Mars had cost in all about $210 billion and achieved few tangible benefits for mankind. How much better if all that money could have been spent on President Kennedy's second proposal, working out how to make our own world a better place to live in, while at the same time undoing some of the damage we have done to it. His vision for communication satellites, his third proposal, has been more than fulfilled and there is no doubt that the investment in satellite technology has revolutionised our lives. It is weather control which has, by comparison, been virtually ignored over the following five decades. The only substantial advantage we have over the rest of nature is that we are an extremely clever species and have a long history of overcoming apparently insuperable obstacles. Surely the time has come to give it a proper go.

The World Meteorological Organization, established in 1950 as a specialised agency, is the UN's authoritative voice on the state and behaviour of the earth's atmosphere, its interaction with the oceans, the climate it produces and the resulting distribution of water resources. One of its departments, the Atmospheric Research and Environment Programme (AREP), has as one of its stated main purposes 'to promote scientific practices in weather modification research'. Regular conferences are organised, the latest being in Bali in October 2011. Some research is being done but there is clearly scope for much more, and with urgency.

COMMON GOALS

We know very little about the part clouds play in influencing weather. We know they carry moisture and that they can both heat and cool the planet. They play a significant role in maintaining the earth's radiation balance, but whether global heating would result in positive or negative feedback through changes in cloudiness seems still to be a mystery. At present there is nowhere near enough computer power to handle all the physics of cloud formation... and so it is all done on 'informed judgements on models'. We all know how inaccurate weather forecasting is. Surely a concerted effort and investment in this subject could change all this! We are not talking here about the convenience of tourists knowing whether to go to the beach, or even giving assistance to farmers wanting to know when to plant or harvest their crops. We are talking about the survival of the planet.

This is a hugely contentious and controversial subject as well as being an exciting field of enormous scientific challenge, which offers the prospect of immense benefit to society. Sadly, like the global heating debate, and indeed like so many other contentious issues, such as nuclear energy, abortion and euthanasia, it polarises opinion, with proponents and opponents making more and more cogent arguments to put their cases, resulting in even greater division. What a tragedy it would be if our global civilisation were to collapse just at the moment when we had the solutions to controlling the climate and feeding everyone within our grasp, for no better reason than that superstition and prejudice made us too afraid to do the necessary research. Economic stringency may get in the way, too, as it will be expensive. The money would be found quickly enough if things really do start to go catastrophically wrong, but by then it may be too late. International politics will inevitably be a major hurdle. 'Whose hand on the thermostat?' is likely to be a rallying cry for inaction. This is the tragedy of the commons writ large. The fear that, even if there is agreement, everything should be shared fairly, and that there will always be those who try to beat the system and are certainly not prepared to trust everyone else will not do the same. Result: no one gets anything. It is a fine hypothesis and of course it has happened. But it doesn't have to be that way.

There is a desperate need for what has been called a 'global federalism of climate policy', in other words, international climate governance, as suggested by President Kennedy back in 1961. This will be especially necessary if geoengineering technologies are to come into play. One such plan is to capture CO_2 direct from the atmosphere and sequester it in underground reservoirs, perhaps the spaces left by the removal of all the coal, oil and gas we have burned over the last couple of centuries. Enough machines doing this could act as a temperature regulator for the whole planet. We could install giant mirrors to orbit the earth and reflect the sun back on itself; or we could inject aerosols into the stratosphere; or fertilise the oceans with hundreds of tons of iron filings. This last would cause plankton to bloom everywhere and that could be very significant. Plankton sequesters about half of all the CO_2 removed from the atmosphere; the other half is pulled out by land plants, anything from grassland to tropical rain forest. Moreover, it would enrich marine life massively and so increase fish stocks. It's like fertilising the oceans and should be much easier, quicker and cheaper than planting trees.

TAMPERING WITH NATURE?

The collapse of the Maya civilisation over a thousand years ago may be seen as a microcosm of what is happening to the whole world today. It is possible they could have saved themselves and prevented the cataclysm that overwhelmed them if they had looked beyond their gods and their traditional solutions to managing the weather. Instead of making more sacrifices and trusting in the power of their beliefs to bring rain, they could have saved their day by restoring the Petén forest, by living within the limitations of their environment and by managing their microclimate in all the ways available to them. Instead they trusted in their gods and sacrifice, and the Four Horsemen had their way. Now we see them galloping into our world and circling our vulnerable corrals. War is as prevalent as ever and the threat of a major conflict is back; in spite of all the achievements of the green revolution, famine still stalks too many desertified regions; new pestilences, such as AIDS and Asian flu, lurk in dark places and threaten to stretch our medical resources to the limit; death comes to millions and yet the population grows remorselessly.

The comparison with what is happening today can be drawn further. Instead of sacrificing people to appease the gods, our modern solution is to manufacture money. Today, when economic collapse, often with environmental roots, threatens to bring our world down, our latter-day shamans, the guardians of the mysteries of how the financial world works, resort to creating money. We call them bankers and we have placed all our faith in them, allowing them to print untold billions of ultimately valueless paper money to appease our gods. There are many who would say that they have as much blood on their hands as the Maya priests who believed that only real blood would make the gods listen.

Set against all these ideas is the belief that tampering with nature is wrong and that by doing so we somehow diminish ourselves or, worse, commit some form of sacrilege. Religious prejudice against 'interfering in God's works' is as ingrained in Christianity as it is in Islam and it is alarmingly pervasive, reaching from the general population of believers to tycoons of industry and politics. A recent poll found that only 26 per cent of Americans accept evolution, while the Qur'an appears not to support the concept that man evolved from other species. I prefer the Sufi saying 'trust in Allah, but tether your camel first', which implies at least collaborating with nature, if not meddling. Or, perhaps, the Christian saying 'God helps those who help themselves.' The moral high ground on this issue is claimed by so many on religious, environmental and scientific grounds, let alone romantic and sentimental ones. Such attitudes would change rapidly and radically if and when the world started experiencing floods or famines caused by collapsing ecosystems.

For me the question is not whether we should interfere, but

how? Almost everything we do today alters our environment, usually detrimentally, as we pollute the atmosphere, exterminate species, exhaust resources and destroy habitats. What we need to do is interfere intelligently, and not just by attempting to restore what we have wrecked; and this is where we have a huge opportunity. For the first time in the history of this planet a species has evolved with the intelligence to cooperate and innovate and imagine new ways of being: us. We have already reshaped the living environment through our domestication of plants and animals, and we have disrupted key ecological systems through our industrial efforts. It is our duty now to start using our intelligence to start putting things right. We also owe it to the many billions of humans we have allowed to flood the earth to have a good life and I believe we have the skill to do this.

Clouds and cloud systems are complex, highly variable and still poorly understood. They can also be very scary. Anyone who has experienced a really powerful thunderstorm will have shared the primeval terror experienced by our ancestors from the first moment they began to think rationally and not instinctively. It is hard not to believe there is a malignant force behind the deafening claps of thunder. Once, while Louella and I were riding through the mountains of central Albania, we and our horses were caught on a forested ridge by a torrential cloudburst. We turned our backs to the horizontal rain, and tried to keep at least a small part of our bodies, where we clutched our cameras and notebooks, dry under our thin capes. Suddenly, there was a blinding flash of lightning followed without a break by a roll of thunder which almost knocked the horses off their feet. We realised we were in the very epicentre of the thunderstorm, which actually felt as though it were below us, moving along the side of the mountain through the valley beneath. For several minutes, flashes and crashes followed each other with such vehemence that we felt as though the storm was communicating directly with us, battering us deliberately. Anyone brought up to believe in evil spirits and mighty gods who battled in the heavens could have been forgiven for being scared and wondering what they were being punished for. Then, as suddenly, the rain stopped, the sun came out and there was a brilliant rainbow overhead. In the Bible the rainbow represents a covenant between God, the human race and every living creature that the flood which covered the earth will not happen again; nevertheless, the dread of nature lies deep in everyone and warns us not to tamper lightly.

Those who believe we have neither the technology nor the wisdom to engineer the planet and that it would be irresponsible and dangerous to attempt to do so may have a case. The history of man's responsibility in handling major technological breakthroughs has not been good, but, as James Lovelock says, 'We are the intelligent elite among animal life on Earth… [and] it is our duty to survive.'[5] This may be the last chance we have to put things right.

The world already faces many crises: potential economic meltdown and a growing need to bail out failing nations; wars that flare up unexpectedly and are often the result of religious differences, which seem to be becoming greater, rather than us all being more ecumenical; desertification, drought and the famines that come after. In spite of these and the huge drain of assets of the major nations, there are still vast financial and research resources available, which could be used to start managing the climate.

I see geoengineers as the new explorers of the big rock we live on, a lump hurtling through space. We suddenly realised how fragile and unique it is when we saw that photograph,

5. Lovelock (2009).

'The Blue Marble', taken on 7 December 1972 by the crew of the Apollo 17 spacecraft from about 28,000 miles away, only a bit further than the circumference of the earth itself. This may have been, in retrospect, the most useful outcome of the whole exercise, as it changed our perception of our world and made us aware of our own fragility. Modern scientists should be climate mechanics, using the new technology as the cutting edge of scientific research, much of it largely neglected still, but one day maybe to be seen as the process whereby mankind's survival was secured. The great danger is that once catastrophes start to happen it will be too late. All available assets will be devoted to mitigating the effects of whatever occurs, be it oceanic flooding, crop failures or mass migration. There will be no time or money left to find out why it all came about and whether it could have been prevented. Even if there is only an outside chance of a global cataclysm, then surely it is worth devoting resources to looking into the possibilities of managing the weather now. And, of course, one way to save and restore the remaining rainforests would be to make sure that they had plenty of rain!

Part 4

THE PALE HORSE: DEATH

The problem: Global warming and pollution

I detect a sea change in the last couple of years, since I started writing this book. Climate deniers seem to be vanishing, with some extremely notable and powerful exceptions, such as Donald Trump of the US, Jair Bolsonaro of Brazil and Scott Morrison of Australia. My impression is that the world is at last waking up to reality; there is genuine awareness and people are demanding action. Never since Nero fiddled as Rome burned has there been a time when inaction in the face of a serious threat has been less excusable. Then civilisation, in the form of the Roman Empire, was threatened by barbarian hordes. Leadership, self-discipline and a determination to pull together to tackle the unmistakeable problem of the decline and imminent fall of the empire were all that was needed; but people chose to believe that the threat was not real and to immerse themselves in distracting activities like sport and increasingly obsessive entertainment. Ring a bell?

Human modification of the atmosphere has long been recognised as real, leading to a new historical era being suggested: the Anthropocene, 'the period during which human activity has been the dominant influence on climate and the environment' (Oxford Dictionaries). The term was first popularised by Paul Crutzen, a German Nobel Laureate described as an 'atmospheric chemist' and one of the progenitors of the Montreal Protocol through his work on the ozone layer. He suggested that it all started in the eighteenth century. Others have proposed that it began as early as 8,000 years ago, when man first began to have an impact on the environment. Now the environment is shaped by man's activities as we raze tropical forests, scrape the ocean floor, mine deep into the earth and impound gigatons of sediment in ultimately useless dams. Humans have become central to the working of the natural world and the time has come when we should recognise this and do something positive about it. Although the natural release of CO_2 from the environment, supplemented by such things as volcanic eruptions, is many times larger than what we emit, our addition matters disproportionately because it unbalances those natural flows. We are now putting more carbon into the atmosphere than can be taken out and that is why we face all the dangers arising from a warmer climate. The knock-on effects are only now becoming apparent and there is real fear in scientific circles that, as happened regularly during the Holocene era, a time may be approaching when the climate may flip into a new state. Whether hotter or colder, this will not be comfortable for mankind.

It is hardly necessary to outline the looming problems. The news is daily full of new apocalyptic stories and dire forecasts. At least people are now talking about it, but there is general agreement that not nearly enough is being done. The temperature is rising inexorably and we just don't know what will happen if, as seems likely, that rise passes the 1.5C above pre-industrial levels at which the IPCC says it must be kept and approaches 2C. There is a real danger that tipping points will start to be passed, if they haven't been already. Glaciers are melting, as is the Greenland ice cap and parts of Antarctica; rivers are disappearing and mountainous regions are seeing more landslides, as the permafrost that held them together melts away. In Siberia there are already fires. In 2019 an unprecedented number of wildfires broke out all round the Arctic, sending smoke across Eurasia. Mostly in Alaska and Russia, the infernos have collectively released more than 120 million tons of CO_2, the most carbon emitted since satellite monitoring began in the early 2000s. By 2100, sea levels could rise by a metre, displacing 10 per cent of the world's population. By that time it may be too late to prevent global catastrophe and an overwhelming number of scientists are now urging action.

Meanwhile, we are also becoming increasingly aware of the extent to which we have damaged the planet and are continuing to do so by polluting the land and the oceans with unconscionable quantities of materials which will not biodegrade, or only do so very slowly. Plastics are the most visible, but detailed research is revealing the equal danger from invisible nanoparticles.

In 1950 the world produced only 2 million tons of plastic per year. Since then, annual production has increased nearly 200-fold, reaching 381 million tons in 2015. For context, this is roughly equivalent to the mass of two-thirds of the world's human population. In other words, we are moving rapidly towards a time when we will produce the combined body weight of the world's population in plastic *each* year. An estimated 3 per cent, or over 11 million tons, enters the sea, 86 per cent in Asia which runs into the Pacific. Most of it reaches the sea by river, far the worst recorded being the Yangtze, which spews out 333,000 tons of plastic each year. Discarded material from fishing fleets also makes a significant contribution to the waste and is particularly harmful, as much of it floats, entangling fish and sea birds. Half of the material in The Great

Pacific Garbage Patch (GPGP) consists of plastic lines, ropes and fishing nets. The whole patch, which is only the largest of many around the world, is over three times the size of Spain and growing exponentially.[1]

Much of the rest vanishes, dissolving, which of course makes it impossible for wildlife not to ingest, or sinks to the bottom of the oceans, where it accumulates.

1. See https://ourworldindata.org/plastic-pollution

The solution: Geoengineering and global clean-up

It will require a massive global effort involving most countries and people everywhere to take the steps necessary to avoid the global temperature rising past the 2C threshold. Weather manipulation could do it, but I'm afraid in the current state of global politics it will be difficult to achieve.

Most agree that the safest and best way to address climate change is to take action to reduce emissions of greenhouse gases, and lots of the countries which signed up to the 2016 Paris Agreement are attempting to do so, with a wide range of determination and success.

However, geoengineering methods, described by the Royal Society as 'the deliberate large-scale intervention in the Earth's climate system, in order to moderate global warming' are likely to be technically possible soon and should be investigated urgently. Some research has already begun, with a notable increase in interest in the subject. All sorts of solutions are being mooted. The technology is new and the default position for scientists is, quite correctly, to be cautious and examine the costs and environmental impacts rigorously. It will be very expensive, but the potential risks to life on earth are so great that now is the time to devote the maximum resources the nations of the world can afford towards research into the possibilities. Nothing can be more important than that, as our very survival depends on it.

REDUCE

The first, safest and most predictable way to respond to the risks of climate change is to reduce and eventually eliminate human-caused emissions of CO_2 and other greenhouse gases. It starts with us and, maybe just in time, that penny does seem to be dropping. People are talking about it now, almost everywhere, the young especially. Extinction Rebellion has captured the public imagination and demonstrations designed to force governments to take action are breaking out globally.

At the same time, at long last, it is becoming economically sensible to do the right thing. Renewable energy is overtaking that generated by fossil fuel and it is rapidly becoming no more expensive to buy electricity from sustainable sources. Electric cars have reached the point where, if not now then very soon, it will make no sense not to use one and petrol and diesel consumption will plummet.

New buildings are spectacularly better insulated, so that there is much less wastage. Fitted with LED bulbs, energy-efficient appliances and water-efficient fixtures, much less energy is frittered away, meaning much less has to be generated.

Food consumption is changing and there is a growing awareness both of the need to eat less meat, most of which is produced in grossly inefficient ways, and to reduce the obscene amount of perfectly good food being disposed of in the environment daily.

And people are increasingly becoming aware of their carbon footprint and wanting to do something about it, like flying less.

Travel companies have been slow in stepping up to the mark to devise carbon-offset schemes, mostly because their clients have been reluctant to pay even a small amount extra for them. There has also been some criticism of the ethics of paying to have trees planted or distributing efficient cooking stoves or low-energy lightbulbs to developing countries in order to assuage the guilt of making a long-distance flight. George Monbiot notably compared carbon offsets with the ancient Catholic Church's practice of selling indulgences.[1] New and more effective ways of carbon offsetting need to be devised. Meanwhile, we need to do more to reduce our vulnerability to the changes in climate which are coming.

ADAPT

To help society adapt to the changing climate means planning for what is likely to occur and acting to prevent or minimise the damage: building flood defences and raising dykes in anticipation of sea level rises, at the same time as managing existing water resources more efficiently; preparing for future extreme weather events by adapting buildings to withstand them; developing drought-tolerant crops and choosing tree species which are less vulnerable to storms, fires and floods; setting aside land for species to survive in and creating corridors in between, along which they can migrate.

Considerable efforts are being made by many communities globally to reduce and adapt and people are largely now aware of the impact of their carbon footprint. Many are trying to make reductions. But it will probably turn out to be too little, too late.

Reducing the use of fossil fuels drastically is also going to

1. See   https://www.theguardian.com/environment/2006/oct/18/green.guardiansocietysupplement

take time, however rapid the development of renewable technology. If we do find that global heating starts to approach the 2C level, which is widely recognised as leading to dangerous and unacceptable consequences, then there will be calls for very urgent action. In such a scenario, geoengineering may be the only solution with the potential to control what happens if the atmosphere starts to change. It has for a long time been a feared and even despised area of science, but attitudes are beginning to change. Both the British Royal Society and the American National Academy of Sciences have recently issued studies that go some way to legitimising it as a climate-adaptation strategy.[2]

There are two radical, but potentially viable, options to intervene currently being researched and developed. One seeks to offset climate warming by greenhouse gases by increasing the amount of sunlight reflected back to space. The other explores ways to remove CO_2 from the atmosphere by sequestration. Neither is a substitute for reductions and adaptations, but as we enter a period of changing climate dramatically different from any previously experienced in recorded human history, interest in deliberate intervention is growing.

MANAGE

It may be possible to manage solar radiation by reflecting some of the sun's energy back into space. Temperature rise is caused by an increased level of greenhouse gases, which absorb energy from the sun and so raise temperatures. This may be counteracted by what is called albedo modification or solar radiation management (SRM). Various technologies have been proposed

2. See https://royalsociety.org/~/media/Royal_Society_Content/policy/publications/2009/8693.pdf and https://nas-sites.org/americasclimatechoices/other-reports-on-climate-change/climate-intervention-reports

to reflect a very small part of incoming sunlight back into space in the hope that this will counteract anthropogenic changes.

This idea was first mooted by James Lovelock and Andrew Watson, who came up with the model of Daisyworld in 1983. In a very simplified form, it is meant to mimic the way the sun and the earth interact and plantlife maintains the stability of the Gaia hypothesis (that living organisms interact with their non-living surroundings to self-regulate the environment and so contribute to maintaining the conditions for life).

This model illustrates the close relationship between the reflectivity of the surface and temperature. The only two life forms are black daisies and white daisies. White daisies reflect more and so absorb less solar radiation than dark daisies as the earth rotates and the sun becomes more or less hot, and so the earth is kept cooler.[3]

A computer model of Daisyworld

3. See   http://www3.geosc.psu.edu/~dbice/DaveSTELLA/Daisyworld/daisy-world_model.htm for a more detailed explanation.

This led to ideas for offsetting greenhouse warming by designing various ways of either diverting or reducing incoming solar radiation (insolation) by making the earth more reflective (increasing the planet's albedo) and so creating a cooling effect.

Space mirrors were first considered in the 1980s and there have been several projects mooted to build machines in space to counter climate change. One of the first was put forward by James Early in 1989, who proposed constructing a 'space shade' 2,000 km in diameter. He estimated the cost at between $1 trillion and $10 trillion and suggested manufacturing it on the moon using a hundred million tons of lunar glass. Various other ideas have been floated, from fabricating trillions of thin metallic reflecting discs from near-earth asteroids[4] to making 10 trillion similar discs on earth and sending them into space a million at a time for 30 years.[5] More realistically, Paul Crutzen has suggested that one quick way to reduce global temperature would be to emulate what has often happened in the past after a major volcanic eruption. Sulphur released into the stratosphere has the effect of cooling the land below and, following these events, there have sometimes been years when the climate cooled markedly. Crutzen proposed that we could bring about the same result deliberately by releasing large quantities of sulphur from all jet airliners, which fly in the stratosphere. He argues that if this were done carefully, with the right doses being released and maintained for long enough, the planet could be cooled.

This idea is once again being backed by Bill Gates, who, in

4. Colin McInnes, 'Space-based geoengineering: challenges and requirements', 2002, https://pdfs.semanticscholar.org/9c16/6ecb81ac3e6d578d04d80f1605d1aa33ce22.pdf
5. Roger Angel, 'Feasibility of cooling the Earth with a cloud of small spacecraft near the inner Lagrange point (L1)', 2006, https://www.pnas.org/content/103/46/17184

2019, put an initial $3 million into a project called Stratospheric Controlled Perturbation Experiment (SCoPEx). Calcium carbonate dust was to be raised by balloon into the atmosphere twelve miles above the New Mexico Desert and released to monitor the dust's sun-reflecting abilities. It was put on hold because of fears that it might trigger a disastrous series of chain reactions and scientists admit the idea is terrifying but, as one of the Harvard team pioneering the project said, '...so is climate change.'

The Marine Cloud Brightening Project (CBP) is a scientific collaboration looking into the impact of particles (aerosols) on clouds and climate. Ironically, although the CO_2 emissions from large ships around the world was calculated in 2019 at more than 130 million tons[6] and international laws are currently being passed to reduce them, recent CBP research has revealed that cloud brightening due to the aerosol particles in these emissions may have reduced global temperature to as much as 0.25C lower than they would otherwise have been. As Kelly Wanser, the principal director, said in a 2019 TED talk, 'We are seeking slow-moving solutions to a fast-moving problem.'[7] It seems likely that there will be more experiments and these will need major investment, but currently there is virtually none and time is running out. There are also no laws governing SRM and it is urgent that they are formulated before disruptive climate change is under way and leaders of countries are called on to act by whatever means are available.

---

6. See   https://ec.europa.eu/clima/news/commission-publishes-information-co2-emissions-maritime-transport_en
7. Kelly   Wanser,   'Emergency   medicine   for   our   climate   fever', https://www.ted.com/talks/kelly_wanser_emergency_medicine_for_our_climate_fever

REMOVE

The other option, called carbon geoengineering or carbon dioxide removal (CDR), is to attempt to remove carbon dioxide (CO_2) and other greenhouse gases from the atmosphere. This can be done by planting trees on a massive scale, by carbon capture and sequestration, whereby CO_2 would be transported for storage in depleted oil or gas fields, and by various other methods. These include adding nutrients to the oceans to draw down CO_2 from the atmosphere, thus triggering large blooms of algae by 'fertilising' the iron-deficient ocean with iron sulphate. The algal bloom would, in theory, absorb carbon, removing it from the atmosphere and then store it by sinking to the sea floor. Another way, as described earlier, and again being promoted by Bill Gates, could be to suck up biologically rich cold water from the deep oceans. All these options are still currently being hampered by lack of funding. That may change as the sense of crisis deepens.

These techniques are in their infancy and would require implementation on a global scale to have any significant impact. They are the big solutions, which require governments to cooperate to make them work. The reports of the IPCC and international treaties like the 2016 Paris Agreement show that the vast majority of scientists and, I believe, most people, understand the problem and want to do something about it, but governments have different agendas and drag their feet. Unfortunately, the will to do this is only likely to be generated by some appalling man-made global catastrophe – and by then it may be too late to put it right. Much better to restore the planet's natural defence systems which, tragically and blindly, we are still doing our best to destroy. There is a lot that we can all do to stop the rot.

Those who believe that geoengineering is a bad idea tend to

concentrate their concern on its potential dangers. What if we do more harm than good? What if this new power falls into the wrong hands? What are the unexpected consequences likely to be? These are the same arguments, with which we are all so familiar, which have been deployed against the development of the nuclear industry for the last few decades. And they do raise legitimate concerns. Geoengineering would almost certainly change the look of the sky by day and by night. It might become whiter by day, with no more blue skies, we are told, although the compensation would be even better sunsets. Night skies would become dimmer, with fewer stars visible and the Milky Way barely discernible. The cultural implications all over the world could be shattering. When, in 1961, I made my very first Saharan camel journey, through the Tassili n'Ajjer mountains in southern Algeria, I was accompanied by two Tuareg, Hamouk and Mahommed. One night, as I lay on the ground in my sleeping bag gazing up into the incredible bowl of stars which make desert travel such a delight, I was astonished to see a star move. This was no shooting star, nor was it an aeroplane; it was a bright heavenly body progressing in a stately but erratic manner across the night sky. I woke my companions up and asked them what it could be. 'Oh yes,' said Hamouk, 'we have seen it too, and it passes every night. Nothing like it has ever been seen before and we are worried that it is an omen.' It was only when I got home that someone enlightened me that it was the Russian satellite, Sputnik, the first to be launched into space four years before. Its erratic headway was accounted for by the fact that it was in a low earth orbit and therefore subject to heat refraction from the surface of the planet. Untold millions have contemplated the familiar night sky through the ages and many, especially those still living in traditional cultures, still do. Tampering with this most significant part of the natural order is indeed a daunting idea.

Global governance of the atmosphere is something humanity has never tried before and I have seen it compared to the first experiments in democracy in Athens, the establishment of the British parliamentary system or the creation of the UN. However, it may be the only way to solve the problem of climate change – the mother of all solutions. What we have been trying to do for the last few decades is patently not working and we have to find another way. The remaining rainforests continue to be cut down, CO_2 emission levels are not declining as fast as most scientists now believe they need to in order to avoid global heating and the world's population heads remorselessly towards 10 billion. It is the job of all governments to come up with solutions to their countries' problems. There is, arguably, no greater predicament facing us today than the potential implications of global climate change. We need to look at all options.

Tampering with the atmosphere might make the sky white or orange and the clouds yellow and purple, perhaps a bit like that sky I saw when the sun rose at dawn over Calakmul. If that were to be the only outcome, it could be a price worth paying, but the other side effects of such tinkering with nature could be worse. Many scientists are opposed to geoengineering, ridicule proposed solutions and urge the precautionary principle because of the potential risks. They advocate a 'middle course', where many of the known technologies, such as wind and PV power, conservation of existing resources and action such as planting many more trees should be implemented before risking climate management. They are all good things to be doing, but they would have to happen at a rapidly accelerating rate, far faster than at present, to have a significant effect. There are signs that this is beginning to happen. For example, there is every chance that there will be a meteoric rise in the number of ultra-efficient electric cars over the next decade or so, but there are also some dangerous political leaders being elected, notably President Bolsonaro of Brazil. As I write, in August 2019, it is being reported that he has made a mocking response to Angela Merkel following Germany's decision to withhold €35 million in funding for Amazon support projects because of the clear-cutting that has been enabled in the first seven months of his regime. He suggested that she should reforest Germany rather than want Brazil to conserve its tropical rainforests. Soon after, he was forced to eat his words when some of the worst forest fires ever broke out across Amazonas and he had to send in the army and ask for international help.

In a special issue of the *Philosophical Transactions of the Royal Society*, the editors speculated that the day may soon come when geoengineering solutions are universally recognised to be less risky than doing nothing. In the same issue, James Lovelock wrote an article in which he diagnosed the earth as having a fever induced by the parasite *Disseminated primatemia* (the superabundance of humans) and recommended as treatment a low-carbon diet combined with nuclear medicine.[8] He maintains that the global heating we have brought upon ourselves 'might already have moved outside our and the Earth's control', but he also tells us that we can fix it if we try, since we are the brains and nervous system of Gaia and we are now responsible: 'Through Gaia I see science and technology as traits possessed by humans that have the potential for great good and great harm. Because we are part of, and not separate from Gaia, our intelligence is a new capacity and strength for her as well as a new danger.'

8. 'A geophysiologist's thoughts on geoengineering', 29 August 2008, https://doi.org/10.1098/rsta.2008.0135

Lovelock has also said that we have two choices: we can live in equilibrium with the planet as hunter-gatherers, or we can live as a very high-tech civilisation. There is nothing in between. Once, long ago, I lived for a while with the Mentawai people on the then still almost pristine forested island of Siberut, off the coast of the great island of Sumatra. Their intimate knowledge of their tropical world beguiled me and I felt completely comfortable and at ease. Their lives were controlled by rituals which bound them to their habitat and, on examination, made good sense and prevented them from over-exploiting it. Before going hunting a ceremony was held at which an elaborate apology was made to the spirits of the animals due to be hunted and it was explained to them that unfortunately it was necessary because the people were hungry. If this was done properly, they told us, they would find the pig or deer, whose spirit had been warned, waiting for them and not running away.

One evening, during the nightly meeting when the whole clan would gather to discuss the day's events, they asked me if I liked being there and how it compared with my distant home. Full of naive enthusiasm, I declared that their way of life was far superior to mine, so tranquil and pure, so pristine and self-confident. 'Then why don't you stay with us? Why must you go?' they asked. When I stumblingly tried to explain that I, too, had a clan and family and that I could not bring them all to Siberut, they fell about laughing, knowing they had caught me out in foolish hyperbole. Much as we may respect them and even try to help them continue their way of life, if that is what they wish, we cannot ourselves become hunter-gatherers again. Lovelock is right. The only way for us is to use our great technological knowledge to engineer ourselves out of the impending catastrophe we have made for ourselves.

CLEANING UP

Meanwhile, there is much that can be done on a smaller scale. The world needs cleaning up before we make it uninhabitable. A great place to start is the ocean. Here, due to a combination of overfishing and pollution, fish stocks have plummeted almost everywhere. Fish catches have, on the other hand, remained fairly stable because as the traditional coastal fishing grounds were fished out, trawlers penetrated into ever deeper waters. Until quite recently, it was virtually impossible to drop nets deeper than 500 metres. Today they are operating at depths of up to 2,000 metres. As the stocks of traditional species were exhausted, the fishing industry turned to other species. Sometimes these have been given new names to make them more attractive. The 'slimehead' was changed to 'orange roughy', while the fish that used to be called 'all mouth', 'molligut' and 'fishing frog', as well as the 'goosefish' in the US, became the monkfish. This is why the global catch has been reported as remaining fairly constant, although the accuracy of the statistics is questionable. But even if this were true, it does not mean that fish stocks have remained stable and there are indications that they may be on the verge of collapse. One reason is that the ocean is warming and this has led to new toxic organisms and algal blooms that poison fish. Many species are moving out of their natural habitats towards the poles in search of cooler waters. An estimated 70 per cent of fish populations are fully used, overused or in crisis as a result of overfishing in warmer waters and it has even been suggested that if the world continues at its current rate of fishing there will be none left by 2050 and the only fish eaten will be those that have been farmed.

In 1964 I set out from the mouth of the Orinoco to go 6,000 miles (10,000 km) through South America by river to the River

Plate. There is no mention in my diaries of seeing any plastic, nor do I recall coming across any floating in any of the rivers, great and small, which I followed. How different it would be today.

The oceans are being polluted by all manner of things, especially plastic. Broadcaster David Attenborough has highlighted this in his *Blue Planet* series as perhaps the greatest threat facing global wildlife. He cites young albatrosses, gannets and puffins being fed plastic. This crisis is now affecting every part of the globe. Recent research in the Arctic has revealed that melting sea ice is freeing previously frozen rubbish, allowing plastic to wash into the area. More than a ton of plastic is dumped in our oceans every minute and already a third of all fish caught contain plastic particles. We are almost all responsible because we use plastic bags and beauty products, such as face washes and shower scrubs, which contain microbeads. Not only should these and many other disposable items containing plastic be banned and governments encouraged to undertake massive clean-up operations, but each and every one of us can help by doing everything we can to reverse the process. Just imagine if each of the seven and a half billion of us were to be inspired to pick up and dispose of at least one piece of plastic waste every day: what a difference we could make! Daunting though it is, such a massive clean-up is not inconceivable. People need to realise just how grave the threats to our survival and those of so many other life forms on the planet are – and act. It is actually happening. One of the most inspiring manifestations of the rapidly growing awareness of the apocalyptic approach of the Four Horsemen has been the many mass movements globally to clear up our towns, cities and countryside and to stop polluting them in the first place. If everyone were to pitch in, the results could really change things.

So, we finish with where we can all begin. There are many

people who have made a difference. Chico Mendes was a humble rubber gatherer in Brazil who single-handedly ignited an international movement which harnessed the concern many were feeling at the devastation of the Amazonian rainforest in which he lived. He was murdered by cattle ranchers at the age of forty-four, but his life had not been in vain. Wangari Maathai was another eco-champion who founded the Green Belt Movement in Kenya and caused 30 million trees to be planted. She was rewarded with the Nobel Peace Prize. The American senator who failed by a whisker to become president, Al Gore, made a huge contribution in raising public awareness about the threat of climate change with his documentary, *An Inconvenient Truth*. And there have been other successes. The Montreal Protocol was signed by many of the world's nations in 1987 outlawing a series of chemicals, most notably chlorofluorocarbons (CFCs), which had been found to be destroying the earth's protective ozone layer. The ozone hole is now healing, giving a glimmer of hope amidst the general environmental gloom. While all too many species are under threat and becoming extinct daily as rainforests, coral reefs and other specialised habitats are being destroyed, there have been a few success stories, such as with pandas, southern white rhinos, grizzly bears, Arabian oryx and Steller sea lions being taken off the official 'endangered' list; but there are all too few others. With proper resources devoted to research and rescue, there could be so many more.

The threat to rainforests

The first simple and obvious step to tackle global warming is to stop cutting down the rainforests, often called the lungs of the world. I had my first taste of tropical forests in early 1958 when I wandered through South-East Asia and Borneo. In those days the rainforest stretched in an uninterrupted swathe across Borneo, the third biggest island in the world, over the central range of mountains and on 500 km across Kalimantan to the Java Sea. Only clear blue rivers teeming with fish and turtles broke the forest canopy, as they meandered to the coast. Very occasionally, a Dayak longhouse might be glimpsed parallel to a riverbank, a building sometimes as long as a man can hit a golf ball, but blending into the landscape and supporting a completely self-sufficient and usually harmonious lifestyle.

In 1977 and 1978 I lived for fifteen months far up one of those rivers as the leader for the Royal Geographical Society of what became the largest British scientific expedition of its day. A hundred and twenty scientists of all disciplines came to study the richest and, by the time we had finished, best researched rainforest in the world. It was in the newly created largest national park in Sarawak, a place called Mulu. They demonstrated how it all worked, how everything was interrelated in the most bounteous of symbiotic relationships, and they solved the puzzle of how such richness and diversity could exist on soil so poor that it was effectively a desert. They demonstrated through innumerable scientific papers the inestimable value of this environment; how the nutrients upon which it all depended were to be found not in the earth below but in the trees themselves, which absorbed them almost as soon as the leaf litter hit the ground. As one, they declared that this priceless incomparable environment should be protected and that to destroy it would bring disaster. The soils would wash away, the rivers silt up; the biodiversity would break down and myriad species would be lost forever, many before they had even been identified.

Thirty years of campaigning followed, during which the global rainforest movement took off and every schoolchild everywhere learned of the significance of this still little-studied habitat, especially the inaccessible and fascinating canopy, where the majority of life is to be found. But the Malaysian and Indonesian governments paid scant attention and looked the other way as corrupt politicians made their fortunes, and the forests were raped. In 1987, I led a mission to Sarawak on behalf of IUCN (the International Union for Conservation of Nature), Friends of the Earth and Survival International to protest directly to the then minister of both tourism and natural resources, James Wong, who also happened to own the largest timber company in the country, that what they were doing to their forests was unsustainable, potentially catastrophic for both the environment and the inhabitants of the interior and already changing the climate, as the first droughts were occurring. It was during his blustering attack on interfering outsiders who knew nothing of the forests that he made the now famous statement, 'I don't like rain. I play golf!'

OIL PALM

Recently, I flew across the island of Borneo and all I could see for most of the time were regimented oil palm plantations. These have rightly been described as the single greatest threat to the most species on earth. More rainforest in Borneo and Indonesia has been cleared and is still being cleared to create these plantations than even that brought about by the original hunger for timber. Oil palm trees are easy to grow and continue to produce for year after year the palm oil for which we in the West seem to have an insatiable appetite. We use it in a huge number of ways, from soap to cooking oil. It is the ingredient often described on the packet as 'vegetable oil' and most people have no idea of the real cost it represents. It is hugely profitable and received a massive boost with the arrival of the fashion for biofuels. The tragedy is that there is masses of grassland and degraded land in the tropics where oil palm plantations would grow perfectly well, but it is more profitable to clear rainforest, even if that involves bribing officials to release supposedly 'protected' forests, because the harvesting of the standing timber means that the plantation is effectively free.

Palm oil is an extraordinarily productive crop, ten times as efficient as the only viable alternative, coconut oil, at producing the lauric acid which is a critical component of so many cosmetic and cleaning products. It would therefore require ten times as much land to produce the same amount from coconuts. And a large proportion of the production, up to 40 per cent in some areas, comes from smallholders, who would be ruined if it were banned. It is the gratuitous clearing of vast tracts of primary rainforest for its production which is one of the greatest threats to the planet. Several sustainability organisations have been formed as a result of international outrage.

They all prohibit open burning and the clearing of rainforest. The one with the most stringent standards and the only one with international recognition, as it is a stakeholder-based NGO involving everyone from growers to campaigners, is RSPO (the Round Table on Sustainable Palm Oil). Unfortunately, it only accounts for about 20 per cent of the global market and there are a plethora of unprincipled cowboy organisations operating in Indonesia and Malaysia, especially in Sarawak, who are responsible for oil palm's appalling reputation. No company should buy palm oil from a supplier who is not a member of RSPO and many in the West now follow this practice. However, over 85 per cent of the world production of 70 million tons comes from Indonesia and Malaysia and most of that goes to India and China, who do not recognise RSPO certification. This really should be one of the preconditions of those two booming nations gaining the respect of the international community.

Wherever the land visible from the plane's window was not covered in chequerboard rows of identical palm trees, all I could see were logging roads following every contour and even climbing inexorably up the previously inaccessible hills of the interior. Between, I could see wide, muddy rivers, which I knew now contain few fish and whose polluted waters flow far out to sea, contaminating the rich mangrove swamps and destroying the breeding grounds of coastal marine life. Progress of a sort, perhaps, but at what a price. Now only 5 per cent of the rich lowland forest remains, and that is almost all in parks and other protected areas. Some of the vast quantity of priceless hardwood timber which has been ripped out of Borneo in the last forty years now lies, I am told, deep in Yokohama harbour in Japan, a good investment for someone. Much of the rest has vanished, having been made into chopsticks and scaffolding.

The worst thing about oil palm trees is the environmental sterility they cause. According to the IUCN, 'the tropical areas suitable for oil palm plantations are particularly rich in biodiversity. Oil palm development, therefore, has significant negative impacts on global biodiversity, as it often replaces tropical forests and other species-rich habitats. Globally palm oil production is affecting at least 193 threatened species, according to the IUCN Red List of Threatened Species. It has been estimated that oil palm expansion could affect 54% of all threatened mammals and 64% of all threatened birds globally. It also reduces the diversity and abundance of most native species.'[1] In Borneo and Sumatra it has played a major role in the decline in species such as orang-utans and tigers.

Some 10,000 of the estimated 75,000 to 100,000 critically endangered Bornean orang-utans are currently found in areas allocated to oil palm. Every year around 750 to 1,250 of the species are killed during human–orang-utan conflicts, which are often linked to expanding agriculture. A small number of species can benefit from the presence of oil palm plantations, including species of wild pig, rodents and some snakes, but by and large oil palm plantations are sterile places with less than 10 per cent of the wildlife found in the remaining (and fast disappearing) adjoining rainforest.

One day palm oil will no longer be needed – the day will surely come when a cheaper substitute is found, a blight comes along and kills all the trees or something else happens to make it irrelevant. Then attempts will have to be made to nourish the impoverished land back to health. Over decades, this may be possible, but the incredibly rich and diverse life and biodiversity of the forest that was there before will be gone forever.

1. https://www.iucn.org/resources/issues-briefs/palm-oil-and-biodiversity

CORRUPTION

During the forty years following our Mulu expedition, over 90 per cent of the lowland rainforests of Sarawak which we had been studying were felled. In the state of Sarawak, the profits from timber, oil and now oil palm (much of it grown on land illegally sequestered from the indigenous communities) have gone straight into the pockets of political cronies, who have run the country for personal profit and plundered it indiscriminately for the past decades of continuous rule. The once unparalleled tropical rainforests have been denuded and the tribal peoples who depended on the forests for their food are now deprived of their means of subsistence and in several cases starving. The corruption was palpable from the beginning. I remember when we were in Mulu in 1978 reading in the local newspaper that the chief minister had granted another large logging concession and been paid $5 million for doing so. He seemed to be almost admired for this and any suggestion of corruption was met with outrage and the real possibility that one would not be let back into the country.

At last, thanks to the courageous campaigning of a few eco-heroes, the cat is now out of the bag and what Gordon Brown has called 'probably the biggest environmental crime of our times' has been exposed. The extent of the profiteering from illegal logging by the alleged kingpin of the Asian timber mafia, Abdul Taib Mahmud, who, incredibly, is still governor of Sarawak (but hopefully will have been ousted by the time this book is published) and his family is staggering. They have profited by at least $15 billion and created a wasteland. All was exposed in a splendid book, *Money Logging*, by Lukas Straumann. More was to come when the indomitable Clare Rewcastle-Brown and her Sarawak Report almost single-handedly uncovered the profound corruption at the heart of the

Malaysian government in the 1MDB scandal, which brought about the resignation and subsequent prosecution of the prime minister, Najib Razak. Over a billion dollars of unexplained cash had been funnelled into the prime minister's own bank account. At last we can talk about corruption without being banned from Malaysia, but that won't bring back what has been lost.

The green pioneers

I believe in goodness not in gods. And I am impatient with the cynicism, corruption, acquisitiveness and short-term attitudes which prevail today. Posterity is my religion. I believe we owe a responsibility to future generations to leave the world a better place. Life is a wonderful gift which we human beings have been granted through an astonishing and unique set of evolutionary circumstances. We should cherish it, nurture it and strive to use it as a means of creating an even better life for future generations.

Throughout the world there are tribal peoples who have developed extraordinarily close and harmonious relationships with their environments and made use of every ecological niche without destroying them in the process. They can help us find this new way of living. They do not contribute to any of the four problems and their lives accord largely to the principles of the solutions: they do not cut down the forests and they do not use fossil fuels; they do not pollute their environment; they maintain stable populations; they do not kill the microbes they live with by using toxic chemicals. As one of them said: 'If it doesn't last for ever, I don't want it.'

Survival International, the global movement for tribal people which I helped found in 1969, has become the world authority on tribal peoples, upholding their right to decide their own future and helping them to protect their lands and their way of life. Displacing communities in the name of development projects such as dams or mines or, even more shockingly, 'for their own good', as some still argue, is always disastrous. It flouts both tribal peoples' own rights and destroys the environment. And yet it is still happening. After forty years of campaigning, I had thought that the message was getting through and it is deeply depressing to see that, again and again, in spite of every telling argument and scientific study, corruption and short-term greed still triumph over sense.

I am someone who has been at the heart of the environmental debate for fifty years, but it was only recently, when I was approached by people researching the origins of the Green movement in Britain, that I realised just how close to the start I was. The original meetings of the group that was to become Survival International took place in my London flat in 1969. I have always maintained that the only reason I was chairman from the start was because it was my cheap white wine they were drinking! Nicholas Guppy and Francis Huxley wrote the initial letter in the *Sunday Times*, following Norman Lewis's powerful article on the genocide being perpetrated on Brazil's Indians, and concern for tribal people remained the focus for many of those of us, like John Hemming and Audrey Colson, who first met there. But Edward (Teddy) Goldsmith, Peter Bunyard, Michael Allaby, Jean Liedloff and Conrad Gorinsky among others believed that the survival of tribal minorities was closely tied to the environmental crisis and that both issues should be addressed together. After about a year, they broke away to start *The Ecologist* magazine, taking with them Robert Allen, who I had recruited as Survival's first director. At first they were based at the Soil Association, which was contracted to produce the magazine, before they moved their office to Teddy's house at Kew. Meanwhile, several of them came to stay regularly at our farm in Cornwall and, in 1972, Teddy Goldsmith (editor), Robert Allen (deputy editor), Michael Allaby (managing editor) and Peter Bunyard all moved down to Cornwall, where Teddy bought a farm not far from mine.

It was within this circle (we have been described as 'eco-pioneers') that many of the core Green political ideas – of sustainability, the stable society and a steady-state economy – were incubated. And it was there that *A Blueprint for Survival*, which has been described as the first eco-manifesto, was produced. The *Blueprint* was one of the first publications to spell out just what a parlous state the world was in and how many matters urgently needed our attention: population explosion; disruption of ecosystems; a looming energy crisis, due to diminishing finite resources; and man's impact on the climate. It became a bestseller, selling 750,000 copies worldwide, has been translated into sixteen languages, was debated in the House of Commons and served as inspiration for many Green manifestos; and it has given rise to political parties all over the world. These include parties in New Zealand, Tasmania and Alsace and the original Movement for Survival, described as the first ecology movement with political ambitions in the world.

In the UK, the PEOPLE Party, the world's earliest Green party, was formed to campaign in the June 1974 UK general election, at which Teddy was to become one of the first six candidates ever for a Green party, standing in Eye in Suffolk. He toured the constituency with a camel, warning that Suffolk might become a desert if environmental damage was allowed to continue unabated. He only got 0.73 per cent of the vote, but the political Green movement was born.

Teddy Goldsmith and friends on the hustings

It seems therefore that I can legitimately claim to be one of the founders of the Green movement. Since then, I have written lots of books about the destruction I have witnessed of forests and peoples. It is easy to despair of man's stupidity, but there is hope. We are a resilient species and maybe we will learn in time what we must do. Perhaps, for once, we will not allow ourselves to destroy our world through materialism, short-sightedness and overexploitation of our limited resources. Perhaps.

HERE WE GO AGAIN

It is irresponsible of us to destroy our world and deny it to future generations, just as the Maya did. What greater crime could there be than this? The Maya may have suspected the cause and effect of their actions and I am sure that towards the end of their era there must have been some who wondered if they had brought it all on themselves. Not by offending their gods and failing to pay them sufficient respect with more and more human sacrifices; but by overexploiting their natural resources. In mitigation, it must be said that they did not know for sure. They didn't have irrefutable scientific evidence to show them that their actions were causing the climatic changes which lay at the root of their problems. Successive droughts were causing crops to fail. Hunger and shortage of land were the cause of wars and famine. Environmental collapse was making the people doubt the infallibility of their gods and kings. They must have wondered.

We have no such excuse. Evidence gathered from the whole planet is available to all. That some choose to ignore it does not alter those facts. The world is in a mess and we as a species are largely responsible. I know, because I have seen it. We have destroyed during my lifetime much of the rainforest that was there at my birth. This is a tragedy. I know also that all too many of the coral reefs I first snorkelled over in the early fifties are now bleached and dying or dead and that what has been described as a cascade of extinctions is taking place in them now, as innumerable often undiscovered species die out. I have seen the pollution of the oceans, which none can deny is caused by us. I fear for the future because I know that my grand-children and great-grandchildren will never enjoy many of the things I have enjoyed. I sincerely hope that I am wrong to feel such urgency, but I cannot believe that striving to put things right can be mistaken. Which is why the articulate naysayers who cloud the issue through the media and the persuasive pro-moters of 'market forces' who encourage ever greater and more profligate consumption of the world's limited resources disturb me so much.

The world, my world, your world, has changed more in my lifetime than for the last 10,000 years, since the end of the last ice age. There is no doubt that we are at least partially responsible. What are we going to do about it? Well, for a start we can look at history to see if there are lessons to be learned from the past.

The seminal moment for me, when my certainty grew that action to save the planet was not just desirable but desperately urgent, was when I revisited the island of Borneo twenty years after living deep in the interior for fifteen months on the Mulu expedition. It was in 1998 and I was appalled by what I saw. That was an El Niño year, the weather system named after the Christ child, since the warm currents which cause it have tended to arrive off the Peruvian coast around Christmas. The climatic effects of El Niño have long been known: floods and droughts and extreme weather generated in different parts of the Pacific area, and indeed much further afield, often far removed from the source. That year was exceptional by any standards and no one was any longer in doubt that much of the increase in problems was being caused by the runaway deforestation in South-East Asia. One simple truth is that primary forest, like that which covered most of the region when I first visited, and which was still dominant right up until the late 1970s, does not normally burn. Regrowth, the vegetation left after the largest trees have been extracted, let alone the secondary scrub resulting from clear felling, does. The extreme drought caused by the 1998 El Niño triggered some of the worst forest fires the world has ever known. Smog blanketed much of the region and major cities, like Singapore, were almost brought to a standstill, with air pollution far above acceptable levels.

I spent the month of June travelling all over the great island of Borneo meeting old friends and going far up some of the innumerable rivers which flow out of the interior. I did not on that occasion go back to Mulu, the place that meant so much to me and held so many memories. I was not ready then. Instead, I went all over that great island to see how things had changed. People I met told me alarming stories. In Sabah, the northern-most part of the island, called North Borneo before it joined the Malaysian Federation in 1963, a senior forester told me that only 5 per cent of the superb old-growth lowland forest now remained. This virgin, primeval forest is uniquely valuable eco-logically, having reached a great age without disturbance. In the Sarawak longhouses I heard stories of how the Dayak peo-ples' traditional forests had been plundered by Japanese logging companies, hand in glove with corrupt Malaysian and Chi-nese officials. I heard how the gentle nomadic Penan, includ-ing some of my good friends from Mulu days, had been driven to putting up barriers across the logging roads that were ruth-lessly, and often illegally, being bulldozed into their remote wilderness. And how some of them had been gaoled and on occasion killed for daring to stand up for their rights.

FIGS AND WASPS

In Brunei I had lunch with a professor at the superb university, funded by the Sultan from his vast oil wealth. He gave me an example of the profoundly alarming potential effects of El Niño. It is well known that wild figs form a vital part of the diet of a great many creatures of the forest. Researchers study-ing birds, mammals and insects know that a fruiting fig, usually a strangling fig, which has taken over some forest giant, is an irresistible magnet. As long as the fig is fruiting it becomes the hub of life as a profusion of creatures compete noisily for the nutritious meal. Figs fruit intermittently throughout the year and so hornbills and macaques, bats and butterflies and lots of other species spend much of their time searching out the latest bonanza and then feasting on it until it is exhausted, when the search begins again. This professor's students had established that the most important variety of fig was propagated by a single type of wasp, the *Agaonidae*, which only bred when conditions were just right, and this was largely dependent on temperature. For millennia, the relatively constant microclimate created by the mighty rainforests of the island of Borneo had provided ideal conditions for the breeding of this wasp and the concomitant propagating of the figs to be triggered intermittently throughout the year. There might be lean times, when everyone had to get by for a while on whatever other food sources could be found, but a bountiful fruiting of a fig was sure to come around before long and a mass migration to the favoured tree would take place, to be followed by a mass gorging. This would trigger in its turn a breeding cycle of fig and wasp and so one more of the complicated symbiotic relationships of the rainforests would be fulfilled.

The new and unprecedented El Niño, the professor had found to his great alarm, appeared to have broken this harmonious and virtuous cycle for the first time in evolutionary history. The exceptional rise in temperature across the island had caused all the fig trees to fruit at the same time. Great while it lasted, but catastrophic for the future. The propagating wasps had all dutifully bred as the fruit ripened, but now there was evidence that the cycle had been broken and no more breeding was taking place. As a result, he feared that this insignificant but absolutely vital wasp was about to become extinct. Without it, the figs would inexorably become extinct, too, since their unique and irreplaceable propagator had gone. The effect might not show for a while, as the figs would go on fruiting, but the implications of them gradually dying out did not bear thinking about. The professor and his students were attempting to breed a colony of the wasps in their laboratory, but he was a worried man.

SUSTAINABLE FORESTRY MANAGEMENT

It was when I reached Kutai, the largest national park in Indonesia, far to the south in Kalimantan, that the extent of the calamity really hit me. This legendary park, last refuge of some of Borneo's most iconic animals, such as orang-utans, sun bears, banteng, gibbons, otters and clouded leopards, had been burned to the ground. As far as I could see, in all directions everything was black and still smoking. In one great fire, this huge park had been virtually destroyed. In the process more CO_2 had been released into the atmosphere in the preceding month than was produced in the whole of Europe in a year. This formed a substantial part of the smog cloud darkening Singapore 1,500 km to the north. There before me was the evidence of something of which I had been dimly becoming aware: that the burning of the world's forests contributed as much CO_2 to global warming as industry. How could this be? And what caused it?

A frightened park guard told me, nervously looking over his shoulder in case his boss should hear him, 'It's all about money. The local timber barons are finding it increasingly difficult to secure logging concessions on virgin forest where, of course, the best trees are to be found. And so they are turning to oil palm plantations, which are becoming very valuable. But they are not allowed just to create a plantation. The land has to be shown to be no good for anything else. And they are supposed to replant all the land they have already logged. So what they are doing is to plant fast-growing timber crops like *Eucalyp-*

tus [gum trees from Australia]. Once partly grown, these burn fiercely, and not just themselves but also the heat sets fire to everything around and so the fire spreads for huge distances. This is what has happened here. There have been so many fires, usually set deliberately, all round the park that our forests have themselves caught fire and everything has been destroyed. They don't care, because now they will be able to say that the land is useless, except to grow oil palm, which will make them rich.'

When I returned home, I found that the world was just beginning to wake up to the gravity of the situation; that I was not alone in believing that combating global warming and climate change should be everyone's top priority. A succession of international conferences was in train, starting with the Kyoto Protocol of 1997, which established greenhouse emissions targets for developed countries and emissions trading of greenhouse gases. It was the first international treaty to combat climate change and, for all its modest targets and failure to bring some major players like the US on board, it promised so much. There was even a global success story to help make us believe that it could achieve its aims. During the 1960s the big issue was acid rain, which was making forests, especially in Canada and Scandinavia, die back and lose their pine needles, while their associated lakes became sterile and devoid of fish. The cause of acid rain was traced to coal-burning power stations, which emitted quantities of sulphur dioxide. Once 'scrubbers' were made sufficiently attractive to polluters, thanks to legal penalties being imposed on them and financial incentives being created through a market in credits, the air cleared up and acid rain is no longer a major problem. This showed what could be done and there seemed no reason why we should not be able to do the same with CO_2.

For the next decade or so, I thought we had the answer to

putting value on the rainforest and so changing things for the better – and I still believe we were right to try. With a couple of friends I helped start a company which attempted to use the terms of the Kyoto Protocol and its subsequent manifestations to value forests, and especially tropical rainforests, through their contribution to reducing CO2 in the atmosphere. By claiming carbon credits for afforestation, reforestation and above all 'avoided deforestation', we hoped to stop the rapid destruction of rainforests and reverse the devastation. And for a time we were very successful, raising many millions of pounds and securing options on the future of many priceless environments which were otherwise threatened. The economic potential of securing carbon credits from replanting degraded rainforests and from avoiding the deforestation of what remained was colossal. The wealth generated was to be put towards funding scientific research into how the workings of rainforests in particular could be better understood and their potential unleashed. For a while things went almost alarmingly well. The company we created became the leading force in acquiring land, concessions and, through innumerable conferences and political lobbying, international action on getting the role of tropical forestry recognised as a priority. Perhaps our high point came when we went to the Bali Climate Conference in 2007. We hosted Lord Stern, author of the Stern Review of 2006, as our keynote speaker and, for a moment, it really looked as though we were going to make a difference. At one point we were told the company was worth $0.5 billion, but of course there was no question of cashing in, as we all believed in the future and that we really were on course to save the world. The global financial turmoil of 2008 undermined the company, which went bust and was wound up, and we all went back to the drawing board, a lot poorer and a little wiser. The full story of those disastrous years may well one day be told, but I have moved on.

OUR RESPONSIBILITIES

Those of us who have been involved for the last few decades in the debate on the global environment have often been accused of being Jeremiahs. I remember my late, great friend Teddy Goldsmith, founder of *The Ecologist*, being interviewed on the BBC in the early seventies. 'What a pessimist you are!' the interviewer said. 'On the contrary,' retorted Teddy, 'You are the pessimist because you agree with all we are discussing about the looming disasters, such as the inevitability of fuel shortages arising due to the finite amount of fossil fuels, but you don't believe anything can be done about it. I am an optimist because I believe that if we act now we can avert many of them. By radically altering our lifestyle we can reduce our consumption of finite resources and we can still save the world.' They laughed at Teddy but they are not laughing now. If, in 1972, they had listened and, as *A Blueprint for Survival* suggested, taken action to mitigate the effect of unsustainable developmental growth, then things would be looking a lot different now.

I believe there is still time, but it is running out. It is time to stop arguing about who and what is to blame and to face up to our responsibilities. And we are quite good at doing this, when it comes to the crunch, although the way is often rocky and nothing comes easy. Over the last couple of millennia we have made astounding technological advances, which have allowed us to foster a wide variety of sustainable lifestyles and create great art. However, these are not universal truths, even within the very limited horizons of the human race. It frustrated me that my elegant and sensitive Tuareg companions in the desert utterly failed to see any beauty in the prehistoric rock art I sought out so assiduously, and laughed uproariously when I played them Mozart.

Only in the late twentieth century did the awareness of human responsibilities to the rest of nature, which are so evident among hunter-gatherer societies, begin to enter mainstream society. The IUCN was an early pioneer, being founded in Switzerland in 1948, from where it still does important work all over the world. In 1961, Julian Huxley and Peter Scott, among others, both later to be founder trustees of Survival, started the World Wildlife Fund. But it wasn't until Rachel Carson's seminal book *Silent Spring* was published in 1962 that people started to wake up to the damage we are doing to our environment. In it she described the devastating effect that pesticides were having on the environment and although chemical companies fought back and pilloried her, she was soon proved correct. In time, pesticides like DDT were banned and, for a time, the world became a safer place. International action to raise popular awareness of the potentially dangerous human-induced change to world climate began with the first United Nations Conference on the Environment in Stockholm in 1972. Teddy Goldsmith and I both spoke there, myself about Survival's embryonic work in Brazil, Teddy about the looming first oil crisis, which he foretold so accurately. I remember being heckled by some early deep-green ecologists from Berkeley, California, who thought that the plight of South American Indians was a purely political issue and that helping them was paternalistic. Late one night, Teddy and I even drafted an idea for a world organisation to protect tropical rainforests by compensating developing countries if they elected to save their forests, and I came up with the name: the World Ecological Areas Programme (WEAP), an acronym I had originally conceived on the hovercraft expedition in 1968, even before Survival was founded. Teddy took the idea a long way, involving such luminaries as Peter Scott, Maurice Strong and Paul Ehrlich. The World Bank expressed an interest, as did several members of the Club of Rome, whose report *Limits to Growth*, also published that year, introduced the idea of sustainable development as opposed to constant economic growth. It even suggested that global overshoot and collapse were likely if current policies continued to be pursued. Teddy received enthusiastic support, but in the end the idea was killed as being too ambitious. We were ahead of our time and when, two or three decades later, major global initiatives did get under way, the Kyoto Protocol being the first serious attempt to reconcile attitudes, much of the remaining rainforest which had been intact in the early seventies had already been destroyed.

At the same time, major advances were made in understanding the relationships between plants and animals and the ways in which they interacted to maintain a functioning ecosystem. Some of these new ideas were introduced to modern agriculture and, as people started to worry about what they were eating, organic farming became fashionable. However, the vast majority of food produced in the world now still comes from the monoculture of a very small number of crops and animals. This is dangerous, but it is argued that without modern techniques such as genetic modifications, which produce 'super crops', the world would starve. It is the catch-22 at the heart of the globalisation debate: whether sustainability or growth should come first and whether they are, in any case, compatible.

Testing times lie ahead, whatever happens. We appear to be on a roller coaster which it is going to be very difficult to stop as what have been called the sleeping giants of the earth are awoken: the potential collapse of the Gulf Stream, the great

Atlantic Ocean heat conveyor; the possibility of gigantic burps of methane being expelled from the frozen tundra as it warms (and there are massive fires in Siberia being reported as I write in 2019); the almost inconceivable impacts of increasing volcanic and tectonic activity and the resultant tsunamis. One of the worst spectres on the horizon attributable to global heating is the threat from melting polar ice caps. The latest research shows that this is happening far more quickly than was previously thought, when much reliance was placed on computer modelling. Raw data collected on polar expeditions by people like Pen Hadow demonstrate that the ice is much thinner than expected. The implications of the meltdown accelerating are grave in the extreme. One of the worst effects would come from the fact that polar ice reflects about 80 per cent of solar radiation. If that white layer at the top of the world went, there would be a net gain of 70 per cent in energy absorption, which of course would hasten global heating dramatically.

Worse still, positive feedbacks could start kicking in. These are not well understood, mostly because they haven't happened in recent history. If the polar ice, which reflects sunlight, is replaced by open water, which absorbs it, more ice melts. If the permafrost starts to thaw, methane is released, which accelerates global heating far more than CO_2. There are many other such scenarios which could make the earth's regulatory systems start to break down and lead to tipping points, when things can start to happen fast and it may be too late to do anything about it. Even if there is only the remotest chance of it coming to pass and of our being able to avert it, then we should take it very seriously. I believe that one of the, perhaps subconscious, origins of the numerous voices raised in opposition to the evidence is a reluctance to accept the scale of what might happen.

The coldest, driest, highest continent on earth is the Antarctic. Its land mass is larger than the US and Mexico combined and it contains 70 per cent of the world's fresh water and 90 per cent of its ice. It has warmed rapidly over the last fifty years. If the ice sheets surrounding the coast were to go into sudden meltdown, and there have been indications recently that this will happen eventually, then global sea levels would rise between five and seven metres. Most of the ice in the northern hemisphere is the sea ice around the North Pole. Core samples and temperature measurements taken over the last century or so, and with increasing urgency in recent years, reveal that the thickness of the ice has almost halved and its margins are retreating as it melts. Because it is sea ice and mostly underwater, its melting has no more impact on sea levels than an ice cube melting in a glass. A much greater danger is posed by the Greenland ice cap. It is the second largest ice body in the world, after the Antarctic ice sheet, and contains colossal quantities of fresh water, enough, once again, to raise sea levels globally about seven metres; and it, too, is beginning to melt. The implications of this acceleration, as some recent research indicates, are grave. Quite apart from the catastrophic rise in sea levels which, even if these were of only a metre or two, would devastate most coastal cities – over 10 million people in Bangladesh live less than one metre above sea level – the influx of so much fresh water could, as I have said above, stop the Gulf Stream, upon whose warming current we in the UK depend.

The reality of this was brought home to me vividly when I travelled to Kamchatka in 1990. This long teardrop-shaped peninsula at the very easternmost end of Russia is about the same size as Britain and lies between precisely the same latitudes: where it is anchored to the Russian mainland lies at 60N, the same as the Shetland Islands, off Scotland; and it is 50N at the southern tip, exactly the same as that of Cornwall, where I live. Throughout the whole peninsula there is permafrost just below the surface of the soil. Scrubby trees grow, but agriculture is virtually impossible and only reindeer thrive. In the winter the temperature drops as low as −60C, and during the very brief summer nothing much but berries grow. This is how it would be for us in Britain if the Gulf Stream stopped.

Gradually, as awareness of these apocalyptic possibilities has entered the public consciousness, they have begun to replace the previous fear of nuclear war or other disaster. By the time the Cold War was coming to an end with the collapse of the Soviet Union, the idea of climate change and all that it might imply had taken firm hold in most cultures, although the emphasis, and the blame, has varied widely. The developing world has blamed the rich nations for polluting the world to feed their industrial revolutions; while the West fears the increasing, uncontrolled damage being done to the global environment by the tiger economies of the East. Attitudes began to change noticeably among world leaders in the early 1990s. For the first time the green debate received encouragement from the Soviet bloc, with sympathetic pronouncements from President Gorbachev.

In recent years, it has become fashionable for politicians, especially in the UK, to vie with each other to be greener than their predecessors. Pious aspirations to live more lightly on the earth now transcend political boundaries and the recognition that something must be done to counteract the damage we have done and continue to do to the environment has become pretty well universal. Ed Miliband, when Environment Secretary in 2007, said in a speech he made at the Vatican: 'Climate change is not just an environmental or economic issue, it is a moral and ethical one. It is not just an issue for politicians or businesses; it is an issue for the world's faith communities.' Tony Blair's campaign director, Alistair Campbell, wrote: 'If you want to be frightened about anything, you want to be frightened about the impact of climate change. It's worrying for our generation – it's even more worrying for the generation coming behind.' Speaking as prime minister just after the coalition government was formed in May 2010, David Cameron said he wanted his new administration to be 'the greenest government ever'. Chris Huhne, his Secretary of State for the Environment, said, 'Climate change is in my view, our view, the greatest challenge facing mankind,' and he added that he wanted to go 'further and faster than ever before'. Unfortunately, he spoiled any confidence those brave words might have inspired by effectively destroying the burgeoning PV industry in Britain six months later. There was widespread speculation at the time that he had been 'got at' by the nuclear industry, which ironically was to take its greatest hit for a generation with the catastrophic earthquake at Fukushima in Japan shortly afterwards. Environment ministers in other countries have expressed similar aspirations.

Ends and beginnings

All over the world, ways are being tried to mitigate the disaster so many see looming. Australia has been hit as hard as any-where by global warming and other recent manifestations of climate change. After 150 years of reasonably reliable winter rains, the pattern of rainfall has decreased by about 15 per cent, creating major problems for farmers and meaning that almost every city in the country faces a water crisis. Much of this change has been attributed to the El Niño effect, which may also be responsible for the extraordinarily severe bush fires, cyclones, floods and other catastrophes which have hit Aus-tralia this century. But it is generally agreed that we must carry some of the responsibility for what has happened.

One scheme with which I was briefly involved and which I personally found very exciting concerned mitigating the increasingly fierce bush fires which raged through the northern territories each year. Until they finally moved off their tradi-tional lands in the 1960s, the aborigines had regularly practised 'spring burning'. Whole families would go out and camp in the bush, setting fire to areas of scrub in order to drive out animals and hunt them. The result was not only that the scrub was kept down but also the grass and other plants, which grew rapidly after the fires, were fresher and more nutritious than if they had not been burned. Much more significant was the fact that six months after the end of the long dry weather, when the more extensive 'winter' burn started spontaneously, often caused by lightning, there was still much less scrub and that larger burn could be controlled or died out naturally. After the clans all moved to settlements and, for two or three decades, ceased to manage a spring burn, it was observed from satellite photography that the winter burns were increasing dramatically in ferocity. Fierce and uncontrollable fires now burn each year right across the region, causing terrible and unprecedented damage to the environment. They also cause vast amounts of CO_2 to be released into the atmosphere.

Australia was one of the first countries to institute programmes for rewarding those who took steps to reduce CO_2 and therein lay the exciting scheme. Some of the clans were encouraged with financial incentives to return to their traditional homelands for a while each year and carry out a spring burn. This not only soon brought about a marked decrease in the winter burns, but also introduced many young aborigines to hunting, gathering 'bush tucker' and experiencing life as their ancestors had lived it throughout history. These were skills which were disappearing fast in urban surroundings and learning them gave them self-confidence and respect for their own rich cultural history. I met the remarkable woman in charge of the programme and felt that here, at last, was a project I could have faith in.

When I spent time with the Yanomami in the Amazon rainforest, they told me about a neighbouring clan whose whole *yano* was burned down. A yano is a large communal house, in which up to 250 people may live. Yanomami means 'people of the yano'. In this case the whole group lost everything except some machetes and bowls which had survived the fire and were found in the ashes. Instead of considering themselves refugees and expecting help from outside, they made a temporary camp and set to work to rebuild their house and replace all that they had lost. All the materials they needed were at hand in the forest. Within a month, they had constructed another magnificent yano, long poles towering up to a high roof and completely thatched with palm leaves. They had woven new hammocks from wild cotton, made new baskets from vines, replaced the bows and arrows which had been burned and used them to provide food while the work went on. Life, they told us, was barely interrupted, because everybody could do everything necessary and so between them the job was done in no time. No specialisation here. No waiting until the electrician or plumber could come. What needed making, they made – and I can attest that their lives were as comfortable, interesting and fulfilled as any in the modern world. The point I am making, however, is not only that the hunter-gatherer life is not sustainable for us any more, as I had realised when I stayed with the Mentawai in Siberut, off Sumatra; but more significantly, there is no way modern humans could adapt to it, even if we wanted or were forced to. We have simply lost the skills and become divorced from the natural world.

The scale of the potential disasters facing mankind from climate change, drought and overexploitation of earth's remaining resources is almost unimaginable, but we have to face up to it. Stephen Hawking said shortly before he died: 'Terror only kills hundreds or thousands of people. Global warming could kill millions. We should have a war on global warming rather than the war on terror.' The UK government's then chief scientist, Sir David King, writing in 2004, agreed: 'In my view, climate change is the most severe problem that we are facing today – more serious even than the threat of terrorism.' For me, it is a far, far greater danger; and yet it is regularly reported and discussed in the same breath as such matters as obesity and bird flu. Our industrial civilisation is alarmingly close to destroying its own conditions of existence and the great danger is that this will lead to an environmental and economic holocaust such as the world has never seen.

James Lovelock fears that if we fail to develop the political will to act together there will be a tenfold reduction in the human population by the end of this century. Not a bad thing, some might say, before thinking through the implications. (Hang on. That means only one in ten of my family and friends will survive! Will my children and grandchildren be among them?) We have already changed the world and its climate radically through our profligate actions and the time has come for us to reverse the process before our own actions destroy us. The only way we can do that is by taking control of the weather and, perhaps, of even more natural phenomena. We have to do whatever it takes to try and ensure our civilisation's survival, if only to buy us some time to understand more and, perhaps, one day get it right.

I often fantasised about this during my travels by camel with the Tuareg in the Sahara. My first journeys with them were in the early 1960s, when I went in search of prehistoric rock art in the Tassili n'Ajjer, Tibesti and Aïr mountains deep in the heart of the desert. In those days, things were very different from how they are today. Reliable sources of water could be found throughout the region, wells were used regularly by the many long-distance caravans criss-crossing the desert carrying salt from the mines or trade goods and there were even some lakes lingering in the rocky fastnesses, one reputed to still have a crocodile or two, isolated for hundreds or even thousands of years from their nearest relations. Although always among the harshest of climates, where thirst was a constant preoccupation, those who were well adapted to the rhythms of desert life, like the Tuareg and also the Toubou to the south and east, survived well and had a vibrant and satisfying culture. Rain was rare but reasonably regular, so that the men could travel great distances and leave their women and children to harvest sufficient crops and rear enough sheep and goats to prosper until their husbands returned, possibly even bearing gifts and occasional luxuries. Then, during the sixties and across the whole region, from the Atlantic coast to the Red Sea, from Algeria to Nigeria and from Libya down to Chad, rainfall virtually ceased. This became known as the Sahel drought, the Sahel being defined as the zone between the Sahara Desert and the savannahs to the south, a region of semi-desert. But this was not a normal drought and it affected a vast area, which included most of the true Sahara Desert.

The crisis was not caused by a simple failure of predicted rains, the sort of droughts to which the Tuareg were well accustomed and with which they were well able to cope. Nor, in this case, was it caused by the people themselves through overgrazing and deforestation. It has been demonstrated by climatologists that increased greenhouse gases caused the sea-surface temperature of the Indian Ocean to rise, weakening the Sahelian monsoon and triggering a dramatic climate shift. The catastrophe was the result of climate change and one of the first pieces of evidence that something terrible and permanent was happening, not just in the Sahel but also in many other parts of the world. In the Sahara, thousands of sheep, goats and camels died and much of the traditional Tuareg way of life collapsed. They migrated in their millions from their homelands in the centre of the Sahara to cities as far distant as Marrakesh and Tripoli, Lagos and Khartoum; and they were settled in refugee camps, where many died. It was a terrible time, with mass starvation leading to local conflicts.

At last, peace of a sort broke out and a little rain began to fall again, enough to allow some Tuareg to return to their lands, which had been almost abandoned, and try to take up their previous way of life. Few managed to return fully to true nomadic pastoralism, though many dreamed of doing so one day, as they accepted menial jobs in rare desert towns or worked in Niger's uranium mines or on other government projects. Some found in the nineties that there was a strong market for tourism to the desert by camel or by Jeep and they supplemented their income by taking groups out into the sand dunes and simulating traditional caravans. Sadly, that business, which I encouraged, has largely collapsed due to the renewed breakdown of law and order over much of the region.

However, as they attempted to move into the twenty-first century, some forty years after the Sahel drought first struck, I was able to travel with them again. Then, the Tuareg would tell me how a sudden light shower months or even years before had caused the desert to bloom and how even now they were reaping the benefit from a hidden rock pool which had not yet dried out. If just a single shower were to fall regularly anywhere on their land just once or twice a year, their whole pastoral economy would change beyond recognition. Flocks of sheep and goats could graze again without, for a time, the need to visit wells, which would in the meantime be filled; nomadic migrations could be planned with confidence. Moreover, bushes and then trees would grow and this in its turn would help to transform the landscape and allow back some settled agriculture. Streams might even start to flow again.

Research has begun into machines which could bring this about. Professor Stephen Salter of the University of Edinburgh designed a spinning turbine more than thirty metres tall and mounted on a ship, which blows seawater through a tube and sends it up in a fine mist to create clouds above.[1] Situated off the coast of West Africa, the clouds would naturally drift over the desert, where they would drop their loads. This technology is now being developed in a different and equally important direction with a view to increasing cloud cover, brightening and whitening the clouds through increasing the water droplets and so, by reflecting sunlight away from the earth, cancelling out the temperature rise caused by man-made climate change. This would involve deploying a large fleet of remotely controlled energy-self-sufficient ships, using the wind to propel themselves and spray the droplets of seawater into the air. If things start to go catastrophically wrong fast, this is likely to be one of the first emergency solutions which will be tried. I believe we should consider doing it now for its original purpose.

The idea of 'managing the planet' fills many, including me, with fear. Do we have the intelligence to manage the earth sensibly? Who are we anyway to try and stabilise the climate, when not only are we patently incapable of managing our own local ecosystems, but caused the problem in the first place? Yet that may just be why we have to try. The seductive alternative is to 'leave it to nature', but nature shows man no special favours and the gradual elimination of us, not just as the dominant polluter but as a whole species, would be horrible. I do not want to see that happen to my grandchildren, which is why I think we should at least consider if we are capable of putting right the damage we have done.

WHAT WENT WRONG?

Why did the Maya civilisation collapse? This is a question which has long been debated. Two possible causes seem more probable than the others. One is that all civilisations arise from

1. See   https://www.heraldscotland.com/news/15540151.professor-stephen-salter-meet-the-scot-whos-invented-a-way-to-calm-killer-hurricanes/

an agricultural base and have a life which is largely dependent on the ability of that resource base to support the urban elite. The other is that a civilisation which develops in a rainforest is particularly vulnerable, because, in addition to the inevitable overexploitation of the area's capacity to produce food, the very act of removing the forest will change the climate and compound the problem. In other words, the Maya civilisation collapsed because they cut the Petén rainforest down. This led to overpopulation, overuse of land and drought, which in turn brought about famine, warfare, disease and so, eventually, collapse. There is a saying that God creates droughts and floods, but that man creates famine. Today, as we see the remaining rainforests of the whole world vanishing and scientists tell us that global climate change is linked to this, the Maya story suddenly becomes highly relevant.

This is not a unique story. In fact it is the very opposite: echoes and parallels can be found everywhere we look on the planet where man has forced the amazing ability of the earth to produce food beyond its sustainable limits. Even the most sophisticated societies have, throughout history, pushed the capacity of their environment past the point where the available resources were capable of supporting them. Sooner or later the system breaks down and somehow the fact that we caused it gets overlooked. Often this is because powerful neighbours, who have been waiting in the wings for weakness to be shown, move centre stage. But usually, although by no means always, it is environmental factors which trigger the collapse. The Maya were not invaded; they simply vanished. The same thing happened to other Amerindian societies around the same time. The Moche and Chimu civilisations prospered on the Pacific coast of what is now Peru until prolonged droughts wiped them out. Further north, the Pueblo Indians of modern New Mexico, whose civilisation reached its height not long after the Maya, were virtually wiped out by a great drought in AD 1130.

The trouble in our own time is that the problem is now a global one and climate change, the population explosion, incipient warfare, the desperation of the hungry and pandemics threaten everyone in the world, even those who have avoided the pitfalls of overexploitation. The current threat of a global collapse of civilisation is so much worse than previous collapses, because we are taking so much else down with us.

The only societies which have escaped this seemingly inevitable crisis have been those who chose not to go down the route of settling, building cities and becoming 'civilised': the hunter-gatherers and hunter-gardeners, who still inhabit many of the remoter parts of the world, especially the rainforests; and also those who were fortunate to settle on rich and fertile land, where the deep soil can take repeated cropping without becoming exhausted and giving up its power to produce. Most of our species throughout history has flouted the concept of sustainability. Again and again, since we first started to chop down trees with stone axes and till the soil beneath, our ancestors demanded more from the earth than it could give and so sowed the seeds of each successive collapse of yet another civilisation. We remember with pride the things we created: the buildings, temples, tombs and art. We create myths about our greatness and rightly revel in the beauty of our music, poetry and literature. And who can blame us when we experience the genius of which we are capable? But what profits us if there is no one left to appreciate all that we have created? Does the music of Mozart echo through space? Are the drawings of da Vinci reflected in cosmic clouds? I fear not. Without a suitable habitat all mankind's creativity will vanish and be forgotten.

No one can ignore the devastation we are wreaking on our planet. I have watched in my own lifetime as the pristine and seemingly endless rainforests of Amazonas and Borneo have been reduced to remnant blocks preserved as national parks and indigenous reserves. I have tried in different ways to play my part in bringing this catastrophe to the attention of those who had the power to halt the destruction and find another way to manage their patrimony. Over the past few decades much of the biological diversity of the planet has been restricted to isolated patches of protected forest, wildernesses and reefs. Confined to national parks and biological reserves, the chances of endangered animals and insects, trees and other plants migrating to breed with other survivors diminish, bringing extinction closer. On every continent things can only get worse as our population continues, inexorably, to grow and consume more. We seem unable to face the facts.

The nearest comparison with the great Mayan ruins is probably the ancient city of Angkor in Cambodia. At its height, between the ninth and thirteenth centuries, this was a vast urban complex described as the largest pre-industrial city in the world, extending over more than 10,000 square kilometres. Recent research indicates that here, too, overexploitation of the forested environment, overpopulation, topsoil degradation and erosion led to its collapse. Other examples of civilisations which dramatically collapsed are Pagan in Burma (Myanmar), Anuradhapura and Polonnaruwa in Sri Lanka and Borobudur in Indonesia. But all these, like Angkor, were overrun and sacked by invaders.

What was very unusual, if not unique, about the Maya collapse, was that it was followed by the abandonment of the area in which they lived for a thousand years, which allowed the rainforest to recover and regrow in a way that has not happened on the same scale anywhere else. With the demise of the great cities of the lowland Maya – Palenque, Tikal, Calakmul, Carakol and the many hundreds of other towns and temples, great and small, which once dominated a rich agricultural landscape – the forest took over and began its efficient work of reducing all back to nature. Creepers rapidly crept into every crevice, forcing the great building blocks apart and allowing trees to root. Soon even the highest temples were covered all over with vegetation and gradually starting to crumble. The destruction of the monuments in this way was as effective as any sacking, but unlike the ancient ruins of the Mediterranean and the Middle East, which often stand forlorn in a barren landscape, here the richness of the Maya culture was replaced by the wealth of a vibrant tropical rainforest.

WHERE NOW?

It is irresponsible of us to destroy our world and deny it to future generations, just as the Maya did. They didn't have irrefutable scientific evidence to show them that their actions were causing the climatic changes which lay at the root of their problems. Successive droughts were causing crops to fail. Hunger and shortage of land were the cause of wars and famine. Environmental collapse was making the people doubt the infallibility of their gods and kings. They must have wondered, but they did not know for sure.

We have no such excuse. Evidence gathered from the whole planet is available to all. The world, my world, your world, has changed more in my lifetime than for the last 10,000 years – indeed, since the end of the last ice age. And who can doubt that we are at least partially responsible? The question remains: what are we going to do about it?

As I sat during that dawn on the top of Temple I at Calakmul at the beginning of this story and gazed out over the forest, I imagined that the land of the Maya stretching to the horizon in all directions was the whole known world, as it was for them.

We know now how thin the layer of breathable air around the world is, a skin-deep film in which our life is possible. So thin is this layer that even our highest mountains, like Everest, poke out above it and to reach the top we need oxygen. High above the mist rising from the forest, I fancied I was in thin air a third of the way up the troposphere, where all sustainable life was below me. It was the oxygen being created through photosynthesis by the boundless vegetation which sustained my life and that of all other creatures which needed air. It was not the lack of oxygen resulting from the Maya destroying their forests which brought about the collapse of their world, but the effect of that collapse was to allow the forest to regrow and, at that moment, I felt myself the beneficiary of the clean air welling up around me from the greenery below. Now that the Maya ecosystem had cured itself over the preceding thousand years, how tragic it would be if the whole cycle of destruction were to be allowed to happen again.

But that is just what looks like coming to pass if nothing is done. The large contiguous biospheres in Mexico, Guatemala and Belize, so wisely created to preserve the integrity of the Petén rainforest, are under threat again from population pressure, illegal logging and cattle ranchers. Ways urgently need to be found to make the forest more valuable as a standing asset to be harvested and used, rather than as a quick cash crop. Otherwise, we will be back at square one. Restoration of the chicle harvest and all the other sustainable wealth of the forests to their maximum capacity is one solution. The story of the demise of the Maya world is just one parable of innumerable other collapses, when societies overstretched themselves and by overconsumption became weak and destroyed their environments. We are now in imminent danger of doing the same thing on a global scale. We know what needs to be done and we are clever enough to do it. Let's make it happen.

CONCLUSION

I do not pretend that the solutions I have proposed in this book are the only ones and that there aren't many other ways we can help to save the world for future generations. But if everyone, especially the old and supposedly wise, were to put their minds to considering what is wrong with the world and what we might do about it, then perhaps, just possibly, it might happen. Because if we don't, if we allow the cacophony of disparate, conflicting and often unashamedly self-interested statements bandied about every day by politicians and the media to prevail, then I fear we will sink inexorably into painful oblivion. We, the human race, are the brightest beings ever to exist on the planet. We are infinitely inventive and creative. In spite of all our faults and frailties, I believe that we and our descendants deserve to live; but we must learn to do so in harmony with the rest of nature, at the same time taking seriously our huge responsibility, as the only thinking beings, to make it work.

Brief bibliography

Battarbee, Richard, *Natural Climate Variability and Global Warming*, Wiley-Blackwell, 2008.

Burroughs, William (ed), *Climate into the 21st Century*, Cambridge University Press, 2003.

Coe, Malcom D., *The Maya*, Thames & Hudson, 2005.

Davey, Edward, *Given Half a Chance*, Unbound, 2019.

Diamond, Jared, *Collapse*, Allen Lane, 2005.

Eatherley, Dan, *Invasive Aliens*, William Collins, 2019.

Flannery, Tim, *Now or Never*, Atlantic Monthly Press, 2009.

– *We Are the Weather Makers*, Penguin, 2006.

Fleming, James Rodger, *Fixing the Sky*, Columbia University Press, 2010.

Goodell, Jeff, *How to Cool the Planet*, Mariner Books, 2010.

Graham, Ian, *The Road to Ruins*, University of New Mexico Press, 2010.

Hulme, Mike, *Why We Disagree about Climate Change*, Cambridge University Press, 2009.

Las Casas, Bartolomé de, *A Short Account of the Destruction of the Indies*, Penguin, 1992.

Lovelock, James, *The Revenge of Gaia*, Allen Lane, 2006.

– *The Vanishing Face of Gaia*, Penguin, 2009.

– *Novacene*, Allen Lane, 2019.

Lynas, Mark, *Six Degrees*, Harper Perennial, 2007.

Leakey, Richard, *The Sixth Extinction*, Wiedenfeld & Nicolson, 1996.

Marsh, George Perkins, *Man and Nature: Or, Physical Geography as Modified by Human Action*, New York 1864.

Martin, Simon & Grube, Nikolai, *Maya Kings and Queens*, Thames & Hudson, 2008.

Matthews, Jennifer P., *Chicle*, University of Arizona Press, 2009.

McDonagh, Father Sean, *To Care for the Earth*, Bear & Co, 1986.

Monbiot, George, *Feral: Rewilding the Land, Sea and Human Life*, Penguin 2014

Money, Nicholas, *The Amoeba in the Room*, 2014.

– *The Selfish Ape*, Reaktion, 2019.

Motavalli, Jim, *Feeling the Heat*, Routledge, 2004.

Nations, James D., *The Maya Tropical Forest*, University of Texas Press, 2006.

Quammen, David, *The Song of the Dodo, A Sidelong View of Science and Nature*, Pimlico, 1996.

The Flight of the Iguana, Simon & Schuster, 1988.

Redclift, Michael, *Chewing Gum*, Routledge, 2004.

Rewcastle-Brown, Clare, *The Wolf Catcher*, Lost World Press, 2019

Smit, Sam, *The Serendipity Foundation*, Unbound, 2016

Straumann, Lukas, *Money Logging: On the Trail of the Asian Timber Mafia*, Bergli, 2014.

Tainter, Joseph A., *The Collapse of Complex Societies*, Cambridge University Press, 1988.

Tree, Isabella, *Wilding: The Return of Nature to a British Farm*, Picador Books, 2018.

Watkinson, Matthew, *The Destiny of Species*, Matador, 2009.

Wulf, Andrea, *The Invention of Nature. The Adventures of Alexander von Humboldt, The Lost Hero of Science*, John Murray 2015.

Acknowledgements

Perhaps surprisingly, as he turned this book down, I would like first to thank Colin Ridler, the brilliant editor of my three books on exploration published by Thames & Hudson. His generous advice and critical comments helped me hugely in redesigning it.

In the nine months between having lunch with John Mitchinson, one of the founders of Unbound, at the Travellers Club and the book being printed, I have had unstinting, imaginative and highly professional support from my three Unbound editors: James Attlee, Andrew Chapman and Xander Cansell. This new way of publishing has been an almost unalloyed pleasure and I recommend it to all.

Of course, my greatest thanks must go to the 161 friends who divvied up so quickly to reach Unbound's crowdfunding target in barely a month. Your names are all listed and I do hope you enjoy the book, even if you don't always agree with me!

Picture credits

Gut microbes – Shutterstock

Chacmool – Luis Alberto Lecuna

Henry Moore – Scottish National Museum of Modern Art

Daisyworld – www.gingerbooth.com